SpringerBriefs in Applied Sciences and Technology

Thermal Engineering and Applied Science

Series Editor

Francis A. Kulacki, Department of Mechanical Engineering, University of Minnesota, Minneapolis, MN, USA

More information about this series at http://www.springer.com/series/10305

Sujoy Kumar Saha • Hrishiraj Ranjan
Madhu Sruthi Emani • Anand Kumar Bharti

Heat Transfer Enhancement in Plate and Fin Extended Surfaces

Sujoy Kumar Saha
Mechanical Engineering Department
Indian Institute of Engineering
Science and Technology, Shibpur
Howrah, West Bengal, India

Madhu Sruthi Emani
Mechanical Engineering Department
Indian Institute of Engineering
Science and Technology, Shibpur
Howrah, West Bengal, India

Hrishiraj Ranjan
Mechanical Engineering Department
Indian Institute of Engineering
Science and Technology, Shibpur
Howrah, West Bengal, India

Anand Kumar Bharti
Mechanical Engineering Department
Indian Institute of Engineering
Science and Technology, Shibpur
Howrah, West Bengal, India

ISSN 2191-530X   ISSN 2191-5318   (electronic)
SpringerBriefs in Applied Sciences and Technology
ISSN 2193-2530   ISSN 2193-2549   (electronic)
SpringerBriefs in Thermal Engineering and Applied Science
ISBN 978-3-030-20738-0   ISBN 978-3-030-20736-6   (eBook)
https://doi.org/10.1007/978-3-030-20736-6

This Springer imprint is published by the registered company Springer Nature Switzerland AG
The registered company address is: Gewerbestrasse 11, 6330 Cham, Switzerland

Contents

Nomenclature

A	Total heat transfer surface area (both primary and secondary, if any) on one side of a direct transfer type exchanger; total heat transfer surface area of a regenerator, m^2 or ft^2
A_c	Flow cross-sectional area in minimum flow area, m^2 or ft^2
b	Distance between plates in a plate-fin exchanger, or channel height, m^2 or ft^2
C_D	Drag coefficient, dimensionless
C_p	Specific heat of fluid at constant pressure, J/kg-K or Btu/lbm-°F
d	Diameter of pin fins, m or ft
D	Mass diffusion coefficient, kg/m-s or lbm/ft-s
D_h	Hydraulic diameter, m or ft
D_m	Minor diameter of a flat tube, defined in m or ft
D_{sh}	Hydraulic diameter based on m or ft
f	Fanning friction factor
G	Mass velocity based on the minimum flow area, kg/m-s or lbm/ft-s
H	Louver fin height, corrugation height, m or ft
h	Heat transfer coefficient based on A, W/m^2-K or Btu/hr-ft^2-°F
h	Fin height ($h = b - t$), m or ft
h_{111}	Mass transfer coefficient, kg/m^2-s or lbm/ft^2-s
j	$StPr^{2/3}$, dimensionless
k	Thermal conductivity of fluid, W/m-K or Btu/hr-ft-°F
K	Pressure drop increment to account for flow development, $K(\infty)$ (full entrance region), $K(x)$ (over length x)
L	Fluid flow (core) length on one side of the exchanger, m or ft
L_P	Strip flow length of OSF or louver pitch of louver fin, m or ft
L_h	Louver height m or ft
L_L	Louver length m or ft
L_S	$(L_p/2s)/(2su/v)$, dimensionless
Nu	Nusselt number
P	Fluid pumping power, W or hp

Pr	Prandtl number
P_f	Fin pitch, m or ft
P_W	Axial wave pitch of wavy fin, m or ft
Re_{Dh}	Reynolds number based on hydraulic diameter, $Re0h = D,GI\mu$, dimensionless
$Re_{Dh,tr}$	Transition from laminar to turbulent flow, dimensionless
Re_s	Reynolds number based on S_1, dimensionless
Re_L	Reynolds number based on the interruption length $= GL/\mu$ dimensionless
Sh_L	Sherwood number based on interruption length, hmL_p/D, dimensionless
S_t	Transverse tube pitch, m or ft
St	Stanton number
s	Spacing between two fins ($=Pt - t$), m or ft
t	Fin thickness, m
T_d	Major diameter of a flat tube, defined in Fig. 5.10, m or ft
T_P	Transverse tube pitch of a flat-tube heat exchanger, m or ft
u	Velocity based on $Ac = s(b - t)$, m^2 or ft^2
v	Velocity based on $Ac = (s - t)(b - t)$, m^2 or ft^2
x	$x/(D_h RePr)$, dimensionless
x^o	Cartesian coordinate along the flow direction, m or ft

Greek Symbols

α	Heat transfer coefficient; apex angle of the fin
β	Helix angle
δ	Liquid film thickness
Δp	Pressure drop
ΔT	Temperature difference
ε	Permittivity
μ	Dynamic viscosity
ρ	Density
σ	Surface tension

Subscripts

ave.	Average
ev	Evaporation
in	Inlet
l	Liquid
s	Saturated
sub	Subcooled
v	Vapor

Abbreviation

CHF   Critical heat flux
CNT   Carbon nanotube
EHD   Electrohydrodynamic
ONB   Onset of nucleate boiling

Additional Nomenclature

A	Constant
A_{ch}	Channel cross-sectional area, m^2
A_h	Heated inside area, m^2
a_1–a_4	Constants
C_0	Parameter in empirical correlations.
C_1–C_2	Constants
Ca	Capillary number
Co	Confinement number
C_p	Specific heat capacity at constant pressure (J/kg K)
B	Bubble diameter, channel diameter, or tube diameter, m.
D_e	Equivalent diameter, same as hydraulic diameter, m
D_h	Hydraulic diameter, m
E_1	Parameter
E_2	Parameter
F	Force, N
F^0	Force per unit length, N/m
F_S	Surface tension force
F	Mass flux, $kg/m^2 s$
g	Gravitational acceleration, m/s^2
h	Heat transfer coefficient, $W/m^2 K$
h_{LV}	Latent heat of vaporization, J/kg
Δh_{Sub}	Subcooling enthalpy, J/kg
m	Mass flow rate, kg/s
n	Parameter in empirical correlation
P_V	Vapor pressure, Pa
P_L	Liquid pressure, Pa
Q	Volumetric flow rate, m^3/s
q	Heat flux, W/m^2
q_{CHF}	Heat flux at CHF, W/m^2
q_{CHF0}	CHF based on channel heated inside area for zero inlet subcooling
q_{CHF04}	Parameters in empirical correlations
q_v	Heat flux, W/m^2
r	Radii, m
Re	Reynolds number

Re_{crit}	Critical Reynolds number
S	Slip ratio
T	Temperature, °C
ΔT	Temperature difference, °C
U	Velocity, m/s

Greek symbols

α	Void fraction, or half angle at the corner included by two channel walls
α_{Hom}	Homogeneous void fraction
β	Volume flow fraction
δ	Film thickness
δ_0	Initial film thickness
δ_t	Thermal boundary layer thickness, m
θ	Contact angle, °
θ_1	Contact angle on one channel surface, °
θ_2	Contact angle on the adjoining surface, °
θ_R	Receding contact angle, °
λ	Parameter in empirical correlations
μ	Dynamic viscosity, kg/m s
ρ	Density, kg/m^3
ρ_m	Average density
S	Surface tension, N/m

Subscripts

c	Cavity
CHF	At critical heat flux condition
ctrit	Critical
eq	Equivalent
exit	At the exit section
G	Gaseous phase
I	Inertia
inlet	At the inlet section
L	Liquid phase
M	Evaporation momentum
min	Minimum
max	Maximum
ONB	Onset of nucleate boiling
r	Receding
S	Surface tension
Sat	Saturation

Sub	Subcooling
τ	Shear (viscous)
V	Vapor
wall	Channel wall

Chapter 1
Introduction

This research monograph deals with the enhanced extended surface geometries for the plate-and-fin heat exchanger geometry as shown in Fig. 1.1. Since the plate-and-fin heat exchanger geometry is conventionally used for gas in one stream and liquid in another stream, the extended surface reduces the gas-side thermal impedances, with the liquid-side heat transfer coefficients being much higher. For both side gas streams, conventionally 50–150% improvement in thermohydraulic performance is achieved and substantial heat exchanger size reduction is accomplished. However, for electronic cooling, liquids like water, FC72, and FC77 enhanced surface geometries are used (Brinkman et al. 1988).

Figures 1.2, 1.3, 1.4, and 1.5 show commonly used enhanced surface geometries. These enhanced surfaces are typically used for laminar flow ($500 < Re_{Dh} < 1500$) and these cannot be used for turbulent flow for fan power limitations. Also, surface roughness elements do not provide appreciable enhancement for low Reynolds number flows. Boundary layer separation, fluid mixing by secondary flows, repeated growth, and wake destruction of the boundary layers are the key enhancers in special channel or surface shapes like wavy channels, vortex generators, offset-strip fins, louvered fins, and perforated fins. Taylor-Gortler vortices and longitudinal vortices generated in the flow dynamics wash the downstream surface (Kays and London 1984; Creswick et al. 1964; Tishchenko and Bondarenko 1983).

A huge amount of data of 52 different plate-fin surface geometries and their corresponding $j–f$ versus Re plots were presented by Kays and London (1984). Shah (1975a, b), Liang and Yang (1975), Pucci et al. (1967), and London and Shah (1968) studied offset fins for gases. Offset fins for liquids have been studied by London and Shah (1968), Sparrow et al. (1977), Mochizuki and Yagi (1977), and Dubrovskii and Fedotva (1972). Smith (1972), Wong and Smith (1966), Davenport (1983), Aoki et al. (1989), Fujikake et al. (1983), and Chang and Wang (1996) studied the performance of louvered fins for heat transfer enhancement in plate-fin heat exchangers. The performance of pin fins and wire screens has been studied by Theoclitus (1966), Hamaguchi et al. (1983), and Torikoshi and Kawabata (1989).

© The Author(s), under exclusive license to Springer Nature Switzerland AG 2020
S. K. Saha et al., *Heat Transfer Enhancement in Plate and Fin Extended Surfaces*,
SpringerBriefs in Applied Sciences and Technology,
https://doi.org/10.1007/978-3-030-20736-6_1

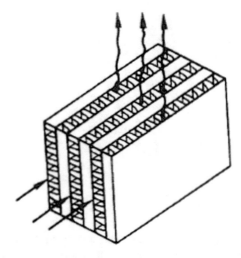

Fig. 1.1 A crossflow plate-and-fin heat exchanger geometry (Webb and Kim 2005)

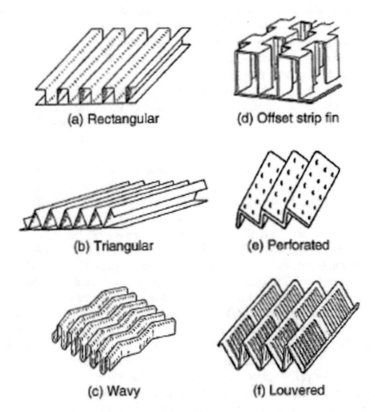

Fig. 1.2 Plate-fin exchanger surface geometries: (**a**) plain rectangular fins, (**b**) plain triangular, (**c**) wavy, (**d**) offset strip, (**e**) perforated, (**f**) louvered (Webb 1987)

Fig. 1.3 Various types of vortex generators. (**a**) Wing-type vortex generators: from left, delta wing, rectangular wing, delta winglet, and rectangular winglet pair; (**b**) illustration of vortices and vortex renewal (Fiebig et al. 1993)

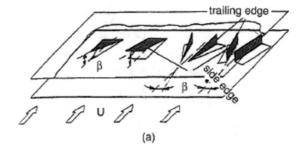

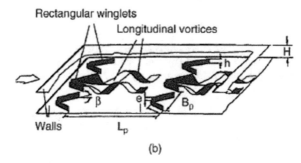

Fig. 1.4 A sketch of longitudinal vortices generated from a winglet vortex generator (Torii et al. 1994)

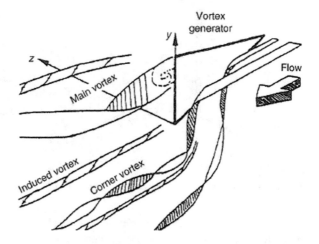

The wavy and herringbone fins for plate-fin heat exchangers have been used by Kays and London (1984), Goldstein and Sparrow (1977), Ali and Ramadhyani (1992), O'Brien and Sparrow (1982), Sparrow and Hossfeld (1984), Molki and Yuen (1986), and Dong et al. (2007). Brockmeier et al. (1993) and Tiggelbeck et al. (1994) worked with vortex generators. Kim et al. (2000), Klett et al. (2000), Bhatt Bhattacharya and Mahajan (2002), Asby et al. (2000), and Calmidi and Mahajan (2000) worked with metal foams. Vasil'ev (2006) and Dubrovsky (1993, 1995) worked on the heat transfer performance of plate-fin interrupted surfaces in channels. Abu Madi et al. (1998), Ayub (2003), Kim et al. (1997), Marr (1990), and

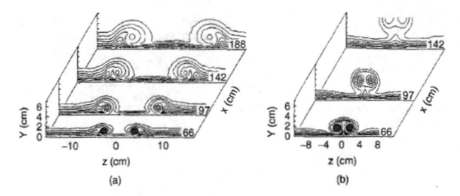

Fig. 1.5 Streamwise velocity contours downstream of two delta winglet vortex generators with (**a**) common-flow-down and (**b**) common-flow-up configuration. Two delta winglet pairs have J 8° angle of attack and 40 mm spacing (Pauley and Eaton 1988)

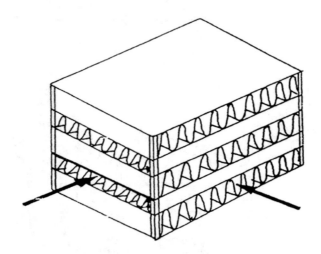

Fig. 1.6 Plate-fin structure for crossflow heat exchanger (Webb 2018)

Wanniarachchi et al. (1995) presented correlations for heat transfer and pressure drop characteristics in plate-fin heat exchangers.

Webb (2018) discussed compact heat exchangers. They are made up of several layers of corrugated plate fins. The compact heat exchangers which are made of aluminum are also called brazed aluminum heat exchangers. They are commonly used in aerospace applications, to separate or liquefy gas, to cool electronic equipment, and in domestic air-conditioners. The plate-fin structure for a crossflow heat exchanger has been shown in Fig. 1.6. The brazed aluminum automotive condenser for condensation of the refrigerant has been shown in Fig. 1.7.

In order to study compact heat exchangers, their geometrical parameters have to be defined. The important parameters are spacing between plate (b), thickness of

Fig. 1.7 Brazed aluminum automotive refrigerant condenser (Webb 2018)

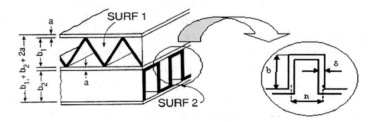

Fig. 1.8 The geometrical parameters of the compact heat exchanger (Webb 2018)

parting sheet (a), fin thickness (δ), pitch of the fin (P_f), surface area per unit volume present between parting sheets (β), ratio of fin surface area, and total surface area (A_f/A). These parameters have been shown in Fig. 1.8. The other derived parameters are as follows:

$$\text{Surface area per unit volume}, \sigma = \frac{\text{Minimum flow area}}{\text{Wetted perimeter}}$$

Surface area per unit volume present between parting sheets,

$$\beta = \frac{\text{Surface area}}{\text{Volume between parting sheets}}$$

Hydraulic diameter, $D_h = \frac{4 \times \text{Minimum flow area}}{\text{Wetted perimeter}}$ and ratio of stream free flow area to frontal area, γ.

Compact heat exchangers for high-temperature solar receivers have been studied by Li et al. (2011). Ariad et al. (1983), Bailey (1997), Edwards et al. (1973), Marsi and Cliffe (1996), Paffenbarger (1990), Parker and Coombs (1980), Jokar et al.

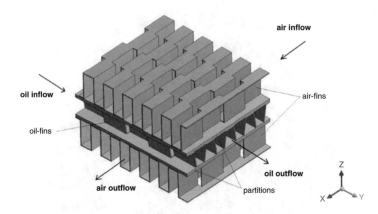

Fig. 1.9 Representation of plate-fin (Zhang et al. 2015)

(2004), Kedzierski (1997, 1998), Lee (1980), and Marvillet (1991) are the others who studied compact heat exchangers.

Zhang et al. (2015) presented the thermal performance of plate-fin heat exchanger. The representation of plate-fin has been shown in Figs. 1.9 and 1.10. The heat transfer coefficient variations with Reynolds number have been shown in Figs. 1.11 and 1.12. Also, the heat transfer enhancement characteristic plot, j/f versus Re, for different fins has been shown in Figs. 1.13 and 1.14. The fin details (fin thickness (δ) and fin offset (d)) have been presented in Table 1.1. The increase in heat transfer coefficient and pressure drop were observed for increase in fin thickness. Also, the j/f factor decreased with increase in fin thickness. They reported a similar trend in heat transfer coefficient, pressure drop, and j/f factor for increase in fin offset. They concluded that the overall performance of fin1 was better than that of fin2 and fin3 and inferior to that of fin4 and fin5.

Wang and Li (2016) discussed the importance of multistream plate-fin heat exchangers especially in cryogenics and presented a detailed review on layer pattern thermal design and optimization. The multistream plate-fin heat exchanger has been shown in Fig. 1.15. Different fin geometries, generally used in plate-fin heat exchangers, have been shown in Fig. 1.16. Table 1.2 shows the commonly used layer pattern units.

Das and Ghosh (2012) presented a review on thermal design of multistream plate-fin heat exchanger. Guo et al. (2015) worked on fin optimization and thermal design of countercurrent plate-fin heat exchangers. Ismail et al. (2010), Wang et al. (2015), Haseler (1983), Prasad (1991), and Wang and Sundén (2001) have worked on multistream plate-fin heat exchangers.

Taler (2004a, b) presented a numerical method for correlations appropriate for the air side using acquired data. The typical equation relating overall heat transfer coefficient U_A to the outer surface A_{rs} of the smooth tube is

$$\frac{1}{U_A} = \frac{A_{rs}}{A_{rin}} \frac{1}{h_c} + \frac{A_{rs}}{A_{rm}} \frac{\delta_t}{k} + \frac{1}{h_o} \tag{1.1}$$

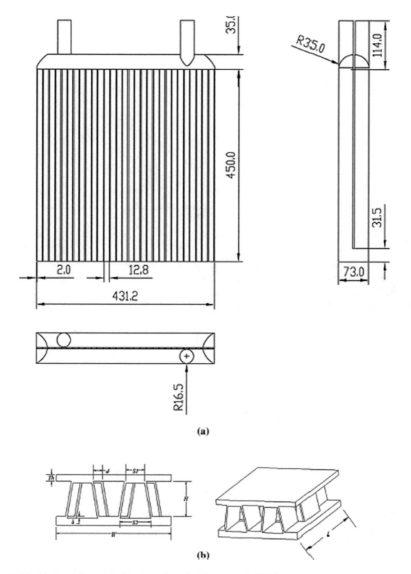

Fig. 1.10 (**a**) Overall model, (**b**) plate-fin unit (Zhang et al. 2015)

$$h_o = h_a \left[\frac{A_{mf}}{A_{rs}} + \frac{A_f}{A_{rs}}(h_a)\eta_f \right]$$ (1.2)

η_f is the ratio of heat flow transferred from physical fins to isothermal fin of the base temperature T_{bf}. Similar work on extended surfaces for heat transfer enhancement has been done by Kays and London (1984), Webb (1994), and Hesselgreaves (2001).

Fig. 1.11 Heat transfer coefficient vs. Reynolds number for fin1, fin2, and fin3 (Zhang et al. 2015)

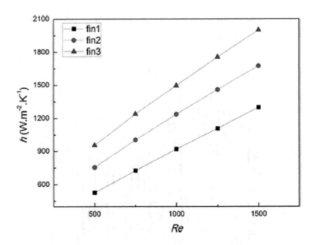

Fig. 1.12 Heat transfer coefficient vs. Reynolds number for fin1, fin3, and fin4 (Zhang et al. 2015)

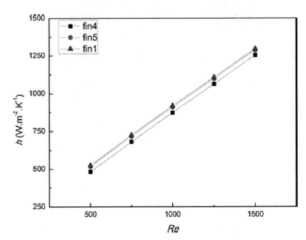

Fig. 1.13 Variation of *j/f* with Reynolds number (Zhang et al. 2015)

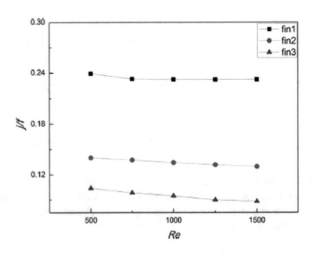

Fig. 1.14 Variation of *j/f* with Reynolds number (Zhang et al. 2015)

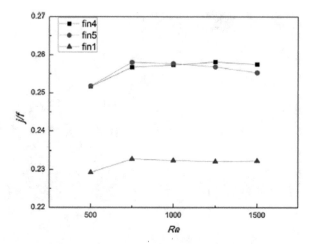

Table 1.1 Details of fin models (Zhang et al. 2015)

Number of case	Parameters	
	δ (mm)	d (mm)
fin1 case	0.1	0.57
fin2 case	0.2	0.57
fin3 case	0.3	0.57
fin4 case	0.1	0.2
fin5 case	0.1	0.4

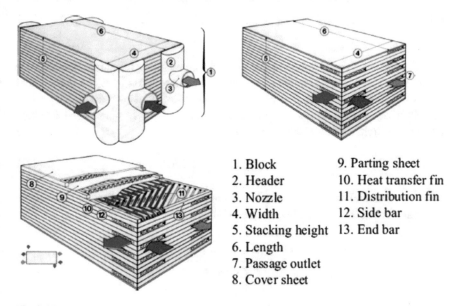

1. Block
2. Header
3. Nozzle
4. Width
5. Stacking height
6. Length
7. Passage outlet
8. Cover sheet
9. Parting sheet
10. Heat transfer fin
11. Distribution fin
12. Side bar
13. End bar

Fig. 1.15 Schematic representation of multistream plate-fin heat exchanger (Wang and Li 2016)

Fig. 1.16 Different fin geometries for plate-fin heat exchangers (Wang and Li 2016)

Taler (2004a, b) used finite volume method for modelling the automotive radiator parameters, Fig. 1.17. He compared the results of Taler (2001) which used analytical model and the results obtained by control volume method for various air frontal velocities ω_o. These results are presented in Table 1.3. He determined the heat transfer coefficient of air side by solving

$$f''_{c,i} - T''_{c,i}\left(h^m_{a,i}\right) = 0, i = 1,2,3,4\ldots n \tag{1.3}$$

for n experimental data sets. Outlet liquid temperature T''_c is the function of air-side heat transfer coefficient h_a. He evaluated Colburn factor and established

$$\frac{Nu_a = c_7 Re_a Pr_a^{\frac{1}{3}}}{1 + c_7 c_8 Re_a} \tag{1.4}$$

Table 1.2 Commonly used layer pattern units (Wang and Li 2016)

Item	Diagrammatize	Note
Single stack		Cold and hot layer alternate arrangement
Double stake		Between the two cold/hot layers is surrounded by two hot/cold layers
Mixed stack		An effective mixed stack of single and double stacking
Temperature crossover		$T_A > T_B > T_C$ and $m_C > m_B$ Owing to improper layer arrangement, the partial cold stream temperature is higher than the hot stream ($T_B > t_A$), the reverse heat transfer against the second law of thermodynamics
Internal heat loss		The heat is not directly transferred from the hot layer to the cold layer, but through the intermediate layer, i.e., the heat transfer between the two cold/hot fluids. $T_A > T_B > T_C$, heat transfer does not allow Q_2 and Q_3 to occur

$$d_{\mathrm{h}} = \frac{4 A_{\min} L_{\mathrm{y}}}{A'_{\mathrm{f}} + A'_{\mathrm{mF}}} \qquad (1.5)$$

where hydraulic diameter on air side manipulates air-side Reynolds number Re_{a}. Further, he experimentally established heat transfer correlations of 18 sets of automotive radiator data via nonlinear regression techniques. These results for automotive radiator are presented in Table 1.3. The heat transfer correlations established according to Table 1.4 are

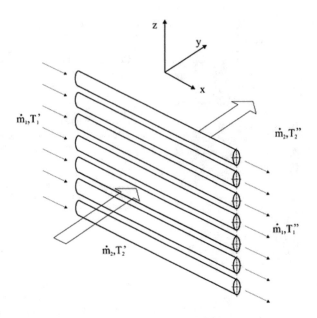

Fig. 1.17 A single-row crossflow heat exchanger (Taler 2004a, b)

Table 1.3 Comparison of results of calculations of radiator using analytical model developed by Taler (2001) and numerical model for various air frontal velocities ω_o (Taler 2004a, b)

No.	ω [m/s]	Analytical model by Taler (2001)			Control volume method		
		$T_c''[^\circ C]$	$T_{cm}[^\circ C]$	$\dot{Q}_{cbl}$ [kW]	$T_c''[^\circ C]$	$T_{cm}[^\circ C]$	$\dot{Q}_{cbl}$ [kW]
1.	0.5	92.29	93.15	5.839	92.26	93.13	5.939
2.	1.0	91.08	92.50	9.768	91.04	92.48	9.885
3.	1.5	90.11	91.97	12.919	90.07	91.96	13.033
4.	2.0	89.28	91.53	15.595	89.24	91.51	15.705
5.	2.5	88.55	91.14	17.940	88.52	91.12	18.045
6.	3.0	87.90	90.78	20.036	87.87	90.77	20.135
7.	3.5	87.32	90.46	21.935	87.29	90.45	22.029
8.	4.0	86.78	90.17	23.673	86.75	90.15	23.763
9.	4.5	86.28	89.90	25.278	86.26	89.88	25.363
10.	5.0	85.82	89.64	26.769	85.80	89.63	26.851
11.	5.5	85.39	89.41	28.163	85.37	89.39	28.241
12.	6.0	84.99	89.18	29.471	84.96	89.17	29.545
13.	6.5	84.61	88.97	30.703	84.58	88.96	30.775
14.	7.0	84.25	88.77	31.868	84.23	88.76	31.937
15.	7.5	83.90	88.58	32.973	83.88	88.57	33.039
16.	8.0	83.58	88.40	34.024	83.56	88.39	34.088
17.	8.5	83.27	88.23	35.026	83.25	88.22	35.087
18.	9.0	82.97	88.06	35.983	82.96	88.05	36.042
19.	9.5	82.69	87.90	36.898	82.67	87.89	36.955
20.	10.0	82.42	87.75	37.776	82.40	87.74	37.831

Table 1.4 Thermal measurement results for automotive radiator (Taler 2004a, b)

No.	ω_o [m/s]	$\dot{V}_c$[L/h]	f'_{am} [$^\circ$C]	f'_c [$^\circ$C]	$f'_c - f''_c$ [K]
1.	1.51	5004	19.5	94.7	2.5
2.	2.50	5009	19.3	94.7	3.7
3.	4.00	5002	19.5	94.7	4.9
4.	5.50	5000	19.7	94.7	5.9
5.	7.00	5004	20.4	94.7	6.5
6.	8.51	5008	20.4	94.7	7.2
7.	1.51	3000	20.1	94.5	4.0
8.	2.51	3006	19.5	94.6	5.9
9.	4.01	3008	19.4	94.6	7.8
10.	5.50	2996	19.3	94.7	9.0
11.	7.00	3005	20.3	94.7	9.9
12.	8.51	2992	20.1	94.7	10.9
13.	1.50	1009	19.8	93.9	10.4
14.	2.51	1000	20.2	94.0	13.9
15.	4.00	996	20.4	94.1	16.4
16.	5.50	991	19.7	94.0	18.9
17.	7.01	1004	20.4	94.1	19.7
18.	8.52	1004	20.3	94.0	21.1

$$j = \frac{9.3802 \times 10^{-3}}{1 + 9.9803 \times 10^{-3} \times 1.2845 \times 10^{-1} Re_a} \tag{1.6}$$

$$j = \frac{9.3802 \times 10^{-3}}{1 + 1.2053 \times 10^{-3} Re_a} \tag{1.7}$$

for $200 \leq Re_a \leq 1500$ and this correction is depicted in Fig. 1.18.

Also,

$$\frac{Nu_a = 9.3802 \times 10^{-3} Re_a Pr^{1/3}}{1 + 1.2053 \times 10^{-3} Re_a} \tag{1.8}$$

for $200 \leq Re_a \leq 1500$.

The numerical simulation showing distribution of the heat transfer coefficients on the finned surfaces has been shown in Fig. 1.19. Also, they plotted Fig. 1.20 for Colburn factor j with Reynolds number from the experimental results and numerical solutions and found that the relative difference is within 16% which is impressive. Finally, he concluded that this method has advantages in evaluating friction and heat transfer characteristics of various types of heat exchangers without involving complex fluid flow field.

Yeh (2001) investigated the one- and two-dimensional heat conduction fin matrix. He determined the optimum aspect ratio of the fins and optimum spacings of longitudinal fin arrays for a given total fin volume, geometry of base plate, and

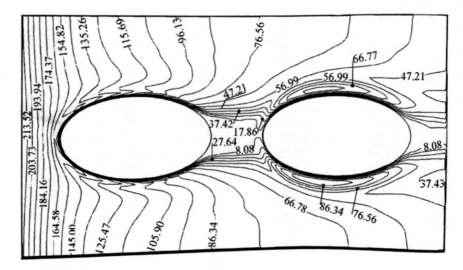

Fig. 1.18 Computed heat transfer coefficient contours on the fin surface (Taler 2004a, b)

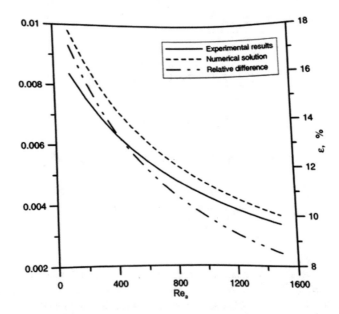

Fig. 1.19 Air-side Colburn j factor as a function of Reynolds number Re_a determined by using present method and numerical simulation (Taler 2004a, b)

Biot number. The heat transfer was considered at the tip of the fin. Two different Biot numbers were taken for one optimum aspect ratio of the fin array. He observed that smaller Biot number should be selected in the design for higher overall surface efficiencies. Kern and Kraus (1972), Kraus (1988), Laor and Kalman (1992),

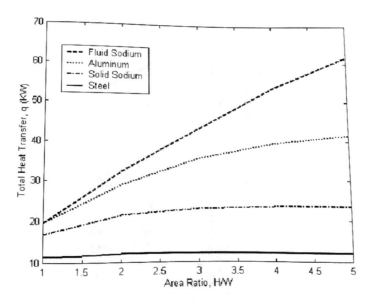

Fig. 1.20 Effect of the area ratio H/W on the rectangular fin heat transfer rate q (Aldoss 2004)

Sohrabpour and Razani (1993), Yeh and Liaw (1993), Chung and Iyer (1993), and Yeh (1994, 1997) intensively studied about the heat transfer characteristics as well as optimum design of extended surfaces with various profiles.

Figure 1.21 shows clearly larger ratio of heat transfer coefficient of fin tip to fin surface (ε) for a smaller aspect ratio at a specified Biot number. Figure 1.22 displays the effect of dimensionless parameter, Biot number, and ε on optimum interspace of the fin in array. Figure 1.23 shows the variation of optimum overall surface efficiency for different Biot numbers and two values of ε. Overall surface efficiencies monotonically decrease with increase in B_i (B_a) as seen from Fig. 1.23. The overall surface efficiency of the finned surface is given by

$$\eta_o \approx \frac{\gamma}{2 + \gamma} \tag{1.9}$$

where γ is the dimensionless parameter WHb/V.

Koyama et al. (1998) studied the numerical simulation of free convective laminar film condensation of pure refrigerant on a vertical surface and compared the heat transfer characteristics of heat transfer in plate-fin condensers. Numerical finite difference method was used to solve the governing equations and boundary conditions in the fin. They examined the effect of fin shape parameters on heat transfer enhancement ratio and fin efficiency. They determined the three-dimension distribution of the fin temperature, distribution of liquid film thickness along the vertical direction, and dimensionless correlation equation for condensation heat transfer. Figures 1.24 and 1.25 show the effect of geometrical parameters of fin on the fin efficiency and average heat flux, respectively.

Fig. 1.21 Dependence of α_o on ε for $B_i = 0.01, 0.1,$ and 0.5 (Yeh 2001)

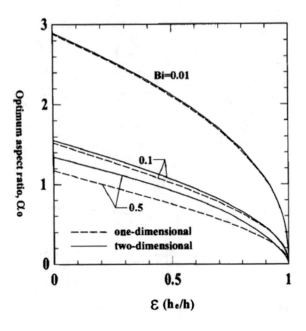

Fig. 1.22 The optimum spacing of fin arrays for $B_i = 0.01$ and 0.1, $\varepsilon = 0$ and 0.5 (Yeh 2001)

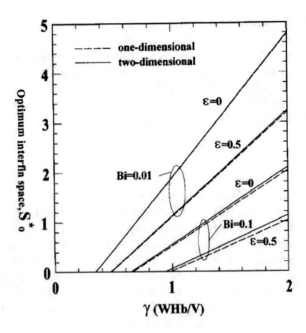

Fig. 1.23 Schematic diagram of overall surface efficiencies of optimum fin arrays (Yeh 2001)

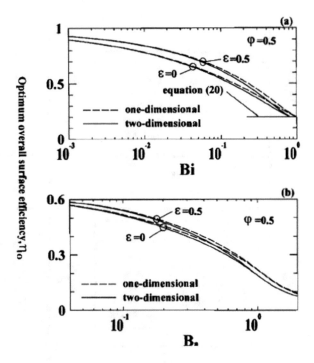

Fin efficiency can be defined as

$$\varphi_m = \frac{\int\limits_0^L \int\limits_0^{S_2} q_w \, ds \, dz}{\int\limits_0^L \int\limits_0^{S_2} q_w^* \, ds \, dz} \tag{1.10}$$

Average heat flux can be defined as

$$q_{wm} = \frac{\int\limits_0^L \int\limits_0^{S_3} q_w \, ds \, dz}{L_p / 2} \tag{1.11}$$

Asano et al. (2004), Bai and Newell (2000), Chiba et al. (1998), Claesson and Simanic (2003), Corberan and Melon (1998), Feldman et al. (2000), Galezha et al. (1976), Hicksen (1999), Hsieh and Lin (2002), Koyama and Tara (1999), Longo et al. (2003, 2004a, b), Mochizuki et al. (1979), Ohadi et al. (1995), Oshima and Iuchi (1974), Ouazia (2001), Panchal et al. (1983), Panitsidis et al. (1975), Peng et al. (1996), Quazia (2001), Raghavan and Murthy (1983), Robertson and Lovegrone (1980), Rovazhyanskiy et al. (1977), Rupani et al. (2002, 2003), Sarma et al. (1989),

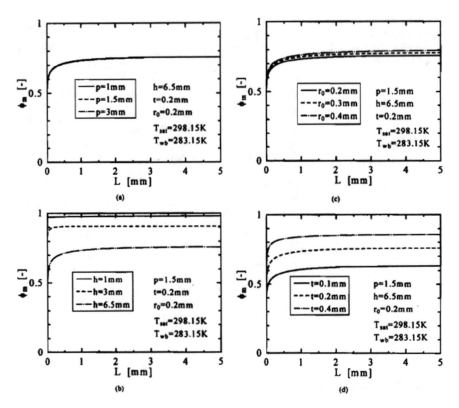

Fig. 1.24 Fin efficiency. (**a**) Effect of fin pitch, (**b**) effect of fin height, (**c**) effect of radius of region II, (**d**) effect of fin thickness (Koyama et al. 1998)

Syed (1991), Vlasogiannis et al. (2002), Wadekar (1989), Wang et al. (1999), Watel and Thonon (2001), Wellsandt and Vamling (2000), Yan and Lin (1999), and Yan et al. (1999) have studied the two-phase flow in a plate-fin heat exchanger.

Aldoss (2004) introduced a unique numerical method for heat transfer enhancement. He used capsulated liquid metal fins and compared the performance from the conventional solid fin. For proper solution, several postulates were identified as (1) infinite depth of fins, (2) no internal heat generation in the flow, (3) negligible viscous dissipation, (4) constant physical properties of liquid metal, and (5) two neighboring fins not affecting one another.

Aldoss (2004) used fluent to solve the thermal performance problem. They compared results of capsulated liquid sodium, solid aluminum, solid sodium, and solid steel fin total heat transfer versus area ratio graph, Fig. 1.26, and found that fluid sodium performance was significantly good. Similarly, he compared the effect of heat transfer coefficient (*h*) in Fig. 1.27 and base temperature in Fig. 1.28. He claimed that capsulated fin heat transfer performance was 500% higher than that of conventional steel fins, 270% higher than that of conventional sodium fin, and 150%

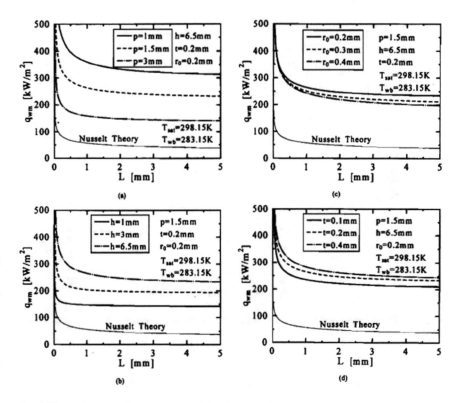

Fig. 1.25 Average heat flux. (**a**) Effect of fin pitch, (**b**) effect of fin height, (**c**) effect of radius of region II, (**d**) effect of fin thickness (Koyama et al. 1998)

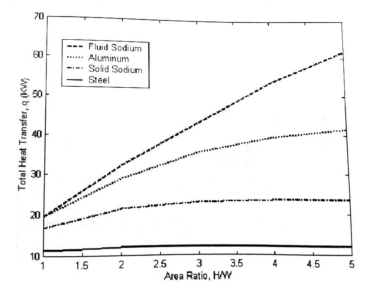

Fig. 1.26 Effect of the area ratio *H/W* on the rectangular fin heat transfer rate *q* (Aldoss 2004)

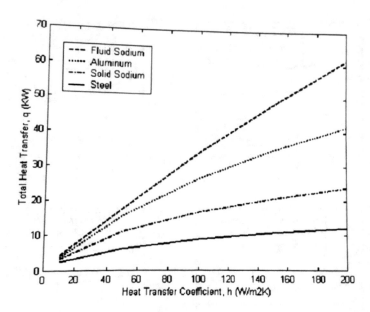

Fig. 1.27 Effect of the convective heat transfer coefficient h on the rectangular fin heat transfer rate q (Aldoss 2004)

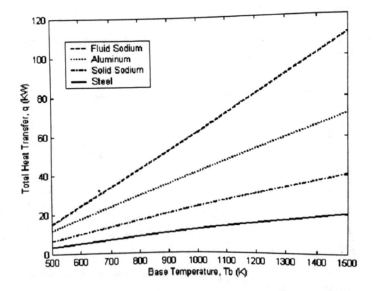

Fig. 1.28 Effect of the base temperature T_b on the rectangular fin heat transfer rate q (Aldoss 2004)

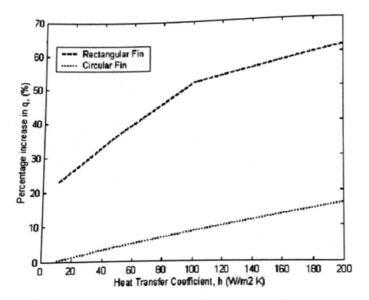

Fig. 1.29 Effect of the convective heat transfer coefficient h on the percentage increase in both rectangular and half-circular fin heat transfer rate q (Aldoss 2004)

higher than that of conventional aluminum fins. Also, performance of rectangular fins was better than that of circular fins as represented in Fig. 1.29.

Gorobets (2008) studied the thermal characteristics of different coated fins and the influence of coating on that fins. The analytical solutions of fins were studied by Barker (1958), Feijoo et al. (1979), Huang and Chang (1980), and Chu et al. (1983) long time ago. The Gorobets study made some additional information in the field of coated fins as he considered composite fins of different shapes and sizes. He developed correlations for uniform coatings. For longitudinal fins, the correlation will be

$$\eta_f = (1 + Bi_c)^{-1th} N_1^- \big/ N_1 \tag{1.12}$$

$$\bar{N}_1^{\,2} = (1 + Bi_c)^{-1} N_f^2, \quad Bi_c = {}^{\alpha \delta_c} \big/ \lambda_c \tag{1.13}$$

For circular fins:

$$\eta_f = 2\left[(X_2^2 - X_1^2)\bar{N}_2 \right]^{-1} \left[AI_1(\bar{N}_2 X_1) - BK_1(N_2 X_1) \right] \tag{1.14}$$

$$B = {}^{I_o}\left(\bar{N}_2 X_1 \right) K_1\left(\bar{N}_2 X_2 \right) \big/ {}_{I_1}\left(\bar{N}_2 X_2 \right) + K_o\left(\bar{N}_2 X_2 \right) \tag{1.15}$$

$$A = {}^{BK_o}\left(\bar{N}_2 X_2 \right) \big/ {}_{I_1 \bar{N}_2 X_2} \tag{1.16}$$

$$\bar{N}_2^{\,2} = (1 + Bi_c)^{-1} N_f^2 \tag{1.17}$$

For cylindrical pin fin:

$$\eta_{\rm f} = \left(^1 + Bi_{c/(1+2\delta_{\rm c})/\delta_{\rm f}}\right)^{-1}{}^{th}\bar{N}\,3\big/\bar{N}_3 \tag{1.18}$$

$$\bar{N}_3^2 = \left[1 + Bi(1 + {}^2\delta_{\rm c}/\delta_{\rm f})^{-1}N_{\rm f}^2\right] \tag{1.19}$$

Similarly, he developed correlations for trapezoidal coating and coating with an arbitrary profile. He also considered the heat transfer problems for composite fins and gave the correlations based on typical assumption. He concluded that thermal efficiency of a longitudinal composite fin can be evaluated by

$$\eta = \frac{1}{h(T_{\rm o} - T_{\rm g})}\int_0^1 \left(T_2\left(x, y = {}^{\delta_1}\!/_{2+\delta_2}\right) - T_{\rm g}\right)\mathrm{d}x \tag{1.20}$$

Similarly, for the composite circular fin

$$\eta = \frac{1}{(x_2^2 - x_1^2)(T_{\rm o} - T_{\rm g})}\int_{x_1}^{x_2} \left(T_2\left(x, y = \frac{\delta_1}{2} + \delta_2\right) - T_{\rm g}\right)x\,\mathrm{d}x \tag{1.21}$$

All the results of this numerical analysis and study are presented in Table 1.5. This table gives a broad picture of the relevancy of this study. Gorobets (2008) study showed that composite fin with coating is more significant than that of fins without coating.

Masri and Cliffe (1996) studied the deposition of fine particles in compact plate-fin heat exchanger. The main purpose of this study was to investigate the performance of plain fin heat exchanger. A simulated model was performed for fouling test with aluminum oxide and ferric oxide particles suspended in water. They used corrugated plain fin and wavy fin. Tests were performed under heating and room-temperature condition. They found the deposition rates under both conditions. They observed that under the heated condition deposition rate was higher.

Under the isothermal condition, mass diffusion was governed due to the effect of Brownian diffusion. The theoretical prediction of mass transfer coefficients from a convective mass transfer correlation was agreed with mass transfer coefficient of experimental data. They showed that pressure drop and deposit weights were higher for wavy fin heat exchanger compared to plain fin heat exchanger.

Pritchard et al. (1992), Eckert and Drake (1959), McNab and Meisen (1973), Bowen and Epstein (1979), Gudmundsson (1981), Turner and Lister (1991), Newson et al. (1983), Sieder and Tate (1936), and Metzner and Friend (1958) explained the mass deposition rate of different types of particles under the flow condition, temperature profile, and pressure profile of fin heat exchangers. Dimensions of the plain and wavy test fins are shown in Table 1.6. The measured and

Table 1.5 Comparison of thermal efficiency for the longitudinal fins with coating calculated in a simplified and two-dimensional model (Gorobets 2008)

$\psi_{Bi} = 0.0025$	$Bi_2 = \frac{ah}{\lambda_2}$	N_f^2	$Bi_c = \frac{a\delta_c}{\lambda_c}$	η_{sim}	η_{two}
$Y_1 = 0.1; Y_2 = 0.12$	25	0.79	0.5	0.689	0.68
	100	1.58	2.0	0.361	0.355
	300	2.7	6.0	0.16	0.157
	1000	5.0	20.0	0.0547	0.0536
	2000	7.07	40.0	0.0284	0.0276
	3500	9.35	70.0	0.0163	0.016
	5000	11.2	100.0	0.0115	0.0112
$Y_1 = 1.0; Y_2 = 1.2$	2.5	0.079	0.5	0.798	0.797
	10	0.15	2.0	0.497	0.506
	30	0.27	6.0	0.248	0.257
	100	0.5	20.0	0.0902	0.0954
	200	0.71	40.0	0.047	0.05
	350	0.94	70.0	0.028	0.0296
	500	1.12	100.0	0.0194	0.021
$Y_1 = 5.0; Y_2 = 6.0$	0.5	0.016	0.5	0.799	0.83
	2	0.032	2.0	0.499	0.574
	6	0.055	6.0	0.249	0.326
	20	0.1	20.0	0.091	0.142
	40	0.141	40.0	0.048	0.081
	70	0.187	70.0	0.028	0.051
	100	0.224	100.0	0.02	0.038

Table 1.6 Shows the dimension of test specimens (Masri and Cliffe 1996)

Fin type	Fin height (mm)	Fin thickness (mm)	Fin pitch (f.p.m.)	a (m²/m)	A_1 (m²/m²)	A_2 (m²/m²)	d_h (mm)
Plain	6.35	0.50	700	0.00379	1.298	8.198	1.59
Wavy	6.35	0.20	710	0.00526	1.712	8.712	2.39

predicted mass transfer coefficients for the plain fin geometry and the wavy fin geometry under the different thermophysical conditions are shown in Tables 1.7 and 1.8, respectively. The variations of pressure drop (ΔP fouled/ΔP clean) versus time for both plain fin and wavy fin are shown in Figs. 1.30 and 1.31, respectively. Bansal and Muller-Steinhagen (1993), Barrow and Sherwin (1994), and Thonon et al. (1999) can be referred for fouling aspects in plate-fin heat exchangers.

Abdel-Kariem and Fletcher (1999), Ay et al. (2002), Berndt and Connel (1978), Charre et al. (2003), Cheng and Xia (2001), Dovic et al. (2002), Dovic and Svaic (2004), Dubrovskii and Fedotva (1972), Edwards (1983), Fletcher and Abdel-Kariem (1999), Fukai and Miyatake (1991a, b), Gut and Pinto (2003, 2004), Jonsson and Moshfegh (2000), Kameoka and Nakamura (1975), Kauzaka et al. (1989), Legkiy et al. (1979), Manglik (1996), Marriott (1977), Mennicke (1972), Muley and Manglik (1997a, b), Okada et al. (1972), Rao and Das (2004), Rosenblad and

Table 1.7 Measured and predicted mass coefficient of plain fin geometry (Masri and Cliffe 1996)

Type of particles (diameter)	Type of test	Re [−]	$Sc \times 10^{-6}$ [−]	Sh [−]	β·calc. $\times 10^6$ (m/s)	β·expt. $\times 10^6$ (m/s)	$(\beta$·expt.$)/$ $(\beta$·calc.$)$
Al_2O_3 (6 μm)	Isothermal	437	8.6	3225	0.189	0.232	1.23
Al_2O_3 (6 μm)	Non-isothermal	533	5.6	3087	0.229	0.400	1.75
Fe_2O_3 (3 μm)	Isothermal	437	4.3	2560	0.299	0.288	0.96
Fe_2O_3 (3 μm)	Non-isothermal	533	2.8	2450	0.362	0.458	1.26

Table 1.8 Measured and predicted mass coefficient of wavy fin geometry (Masri and Cliffe 1996)

Type of particles (diameter)	Type of test	Re [−]	$Sc \times 10^{-6}$ [−]	Sh [−]	β·calc. $\times 10^6$ (m/s)	β·expt. $\times 10^6$ (m/s)	$(\beta$·expt.$)/$ $(\beta$·calc.$)$
Al_2O_3 (6 μm)	Isothermal	603	8.6	3788	0.147	0.281	1.91
Al_2O_3 (6 μm)	Non-isothermal	736	5.6	3628	0.180	0.457	2.54
Fe_2O_3 (3 μm)	Isothermal	603	4.3	3007	0.234	0.371	1.58
Fe_2O_3 (3 μm)	Non-isothermal	736	2.8	2879	0.283	0.532	1.87

Fig. 1.30 Pressure drop results for the plain-fin geometry (Masri and Cliffe 1996)

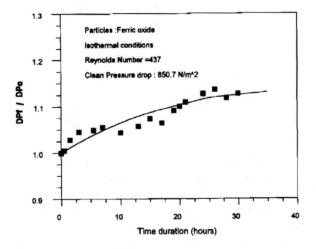

Kullendorf (1975), Sparrow and Liu (1979), Talik et al. (1995a, b), Thonon et al. (1995), and Wang and Sunden (2003) can be referred for various works on plate heat exchangers.

Fig. 1.31 Pressure drop results for the wavy-fin geometry (Masri and Cliffe 1996)

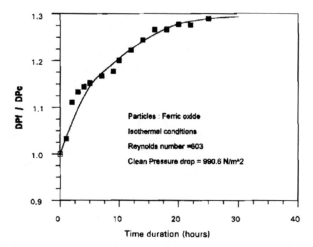

References

Abdel-Kariem AH, Fletcher LS (1999) Comparative analysis of heat transfer and pressure drop in plate heat exchangers. In: Proc of the 5th ASME/JSME thermal eng conf, San Diego, CA, AJTE99-6291

Abu Madi M, Johns RA, Heikal MR (1998) Performance characteristics correlation for round tube and plate finned heat exchangers. Int J Refrig 21:507–517

Aldoss TK (2004) Using capsulated liquid metal fins for heat transfer enhancement. J Enhanc Heat Transf 11(2):151–160

Ali MM, Ramadhyani S (1992) Experiments on convective heat transfer in corrugated channels. Exp Heat Transfer 5:175–193

Aoki H, Shinagawa T, Suga K (1989) An experimental study of the local heat transfer characteristics in automotive louvered fins. Exp Therm Fluid Sci 2:293–300

Araid FF, Awad MM, El-hadik AA (1983) Heat transfer and hydrodynamic resistance in compacting plate heat exchanger type 'divergent-convergent' surfaces. ASME-JSME Therm Eng Joint Conf Proc 3:437–441

Asano H, Takenaka N, Fujii T (2004) Flow characteristics of gas–liquid two- phase flow in plate heat exchanger: (visualization and void fraction measurement by neutron radiography). Exp Therm Fluid Sci 28:223–230

Asby MF, Evans AG, Fleck NA, Gibson LJ, Hatchinson JW, Wadley HNG (2000) Metal foams: a design guide. Butterworth & Heinemann, Boston

Ay H, Jang J, Yeh J-N (2002) Local heat transfer measurements of plate finned-tube heat exchangers by infrared thermography. Int J Heat Mass Transfer 45:4069–4078

Ayub ZH (2003) Plate heat exchanger literature survey and new heat transfer and pressure drop correlations for refrigerant evaporators. Heat Transfer Eng 24(5):3–16

Bai X, Newell TA (2000) Two-phase flow characteristics in Chevron-style flat plate heat exchangers. In: Proc 2000 int refrigeration conf at Purdue, pp 87–94

Bailey KM (1997) Plate heat exchangers: a compact heat exchanger technology. In: Proc int conf on compact heat exchangers for the process. Ind Begell House Inc., New York, pp 1–10

Bansal B, Muller-Steinhagen H (1993) Crystallization fouling in plate heat exchangers. J Heat Transfer 115:584–591

Barker JJ (1958) The efficiency of the composite fins. Nucl Sci Eng 3:300–312

Barrow H, Sherwin K (1994) Theoretical investigation of the effect of fouling on the performance of a tube and plate fin heat exchanger. Heat Recov Syst CHP 14:1–5

Berndt T, Connell JW (1978) Plate heat exchangers for OTEC. In: Proc the fifth ocean thermal energy conversion conf, pp VI-288–VI-320

Bhatt Bhattacharya A, Mahajan RL (2002) Finned metal foam heat sinks for electronics cooling in forced convection. J Electron Packag 124:155–163

Bowen BD, Epstein N (1979) Fine particle deposition in smooth parallel-plate channels. J Colloid Interface Sci 72(1):81–97

Brinkman R, Ramadhyani S, Incropera FP (1988) Enhancement of convective heat transfer from small heat sources to liquid coolants using strip fins. Exp Heat Transfer 1:315–330

Brockmeier U, Guntermann T, Fiebig M (1993) Performance evaluation of a vortex generator heat transfer surface and comparison with different high performance surfaces. Int J Heat Mass Transfer 36:2575–2587

Calmidi VV, Mahajan RL (2000) Forced convection in high porosity metal foams. J Heat Transfer 122:557–565

Chang YJ, Wang CC (1996) Air side performance of brazed aluminum heat exchangers. J Enhanc Heat Transf 3:15–28

Charre O, Jurkowsk R, Bailly A, Maziani S, Altazin M (2003) General model for plate heat exchanger performance prediction. J Enhanc Heat Transf 10:181–186

Cheng L, Xia G (2001) Characteristics of single-phase heat transfer and pressure drop in plate heat exchangers with and without PTFE coatings. In: Exp heat transfer, fluid mechanics, and thermodynamics (2001) Edzioni ETS, Pisa, Italy, pp 1815–1820

Chiba T, Kinoshita T, Shinmura T, Aoki H, Nakajima Y (1998) New development of plate and fin evaporator. In: SAE int congress and exposition SAE 981179

Chu HS, Weng CI, Chen CK (1983) Transient response of a composite straight fin. J Heat Transfer 105(2):307–311

Chung BT, Iyer JR (1993) Optimum design of longitudinal rectangular fins and cylindrical spines with variable heat transfer coefficient. Heat Transfer Eng 14(1):31–42

Claesson J, Simanic B (2003) Pressure drop and visualization of adiabatic R134a two-phase flow inside a Chevron type plate heat exchanger. In: Proc the 21st IIR int congress of refrigeration, Washington, DC, ICR 314

Corberán JM, Melón MG (1998) Modelling of plate finned tube evaporators and condensers working with R134A. Int J Refrig 21:273–284

Creswick FA, Talbert SG, Bloemer JW (1964) Compact heat exchanger study. Battelle Memorial Institute Report, Columbus

Das PK, Ghosh I (2012) Thermal design of multistream plate fin heat exchangers—a state-of-the-art review. Heat Transfer Eng 33(4–5):284–300

Davenport CJ (1983) Heat transfer and flow friction characteristics of louvered heat exchanger surfaces. In: Taborek J, Hewitt GF, Afgan N (eds) Heat exchangers: theory and practice. Hemisphere Publishing, Washington, DC, pp 387–412

Dong J, Chen J, Chen Z, Zhou Y, Zhang W (2007) Heat transfer and pressure drop correlations for the wavy fin and flat tube heat exchangers. Appl Therm Eng 27:2066–2073

Dović D, Palm B, Švaić S (2002) Basic single phase flow phenomena in Chevron type plate heat exchangers. Zero leaks – minimum charge, IIR/IIF, Stockholm, Sweden Paper H4

Dovic D, Svaic S (2004) Experimental and numerical study of the flow and heat transfer in plate heat exchanger channels. In: Tenth international refrigeration and air conditioning conference at Purdue, R097

Dubrovskii EV, Fedotva AL (1972) Investigation of heat exchanger surfaces with plate fins. Heat Transfer Sov Res 4:75–79

Dubrovsky EV (1993) Highly effective plate-fin "heat exchanger surfaces: from conception to manufacturing". In: Aerospace heat exchanger technology, pp 501–547

Dubrovsky EV (1995) Experimental investigation of highly effective plate-fin heat exchanger surfaces. Exp Therm Fluid Sci 10(2):200–220

Eckert ER, Drake G (1959) Heat and mass transfer, 2nd edn. McGraw-Hill, New York

Edwards MF (1983) Heat transfer in plate heat exchangers at low Reynolds numbers in low Reynolds number flow heat exchangers. Hemisphere Publishing Corporation, Washington, DC, pp 933–947

Edwards FJ, Henry TA, Hayward GL (1973) The characteristics of the tube and continuous plate fin type of compact heat exchanger. In: Conf on recent dev in compact high duty heat exchangers, Institution of Mech Eng, pp 53–61

Feijoo L, Davis HT, Ramkrishna D (1979) Heat transfer in composite solids with heat generation. J Heat Transfer 101(1):137–143

Feldman A, Marvillet C, Lebouché M (2000) Nucleate and convective boiling in plate fin heat exchangers. Int J Heat Mass Transfer 43:3433–3442

Fiebig M, Valencia A, Mitra NK (1993) Wing-type vortex generators for fin-and- tube heat exchangers. Exp Therm Fluid Sci 7:287–295

Fletcher LS, Abdel-Kariem AH (1999) A comparative analysis of heat transfer and pressure drop in plate heat exchangers. In: Proc 5th ASME/JSME thermal eng joint conf, Paper ATJE99-6291

Fujikake K, Aoki H, Mitui H (1983) An apparatus for measuring the heat transfer coefficients of finned heat exchangers by use of a transient method. In: Proc of Japan 20th symposium on heat transfer, pp 466–468

Fukai J, Miyatake O (1991a) Laminar-flow heat transfer within parallel-plate channel with staggered baffles. Abstract Kagaku Kogaku Ronbun 17:325

Fukai J, Miyatake O (1991b) Evaluation of heat transfer augmentation for laminar flow within parallel plate channel with staggered baffles. Abstract Kagaku Kogaku Ronbun 17:904

Galezha VB, Usyukin JP, Kan KD (1976) Boiling heat transfer with Freons in finned-plate heat exchangers. Heat Transfer Sov Res 8(3):103–110

Goldstein LJ, Sparrow EM (1977) Heat/mass transfer characteristics for flow in a corrugated wall channel. J Heat Transfer 99:187–195

Gorobets V (2008) Influence of coatings on thermal characteristics and optimum sizes of fins. J Enhanc Heat Transf 15(1):65–80

Gudmundsson JS (1981) Particulater fouling. In: Fouling of heat transfer equipment, pp 357–387

Guo K, Zhang N, Smith R (2015) Optimisation of fin selection and thermal design of counter-current plate-fin heat exchangers. Appl Therm Eng 78:491–499

Gut JAW, Pinto JM (2003) Modeling of plate heat exchangers with generalized configurations. Int J Heat Mass Transfer 46:2571–2585

Gut JAW, Pinto JM (2004) Optimal configuration design for plate heat exchangers. Int J Heat Mass Transfer 47:4833–4848

Hamaguchi K, Takahashi S, Miyabe H (1983) Heat transfer characteristics of a regenerator matrix (case of packed wire gauzes). Trans JSME 49B-445:2001–2009

Haseler L (1983) Performance calculation methods for multistream plate fin heat exchangers. In: Taborek J, Hewitt GF, Afgan N (eds) Heat exchangers—theory and practice. Hemisphere Publishing, New York, pp 495–506

Hesselgreaves JE (2001) Compact heat exchangers: selection. In: Design and operation

Hicksen DC (1999) Boiling and condensation heat transfer coefficients in a plate heat exchanger. In: IMeChE conf trans, 6th UK national conf on heat transfer, pp 133–139

Hsieh YY, Lin TF (2002) Saturated flow boiling heat transfer and pressure drop of refrigerant R-410A in a vertical plate heat exchanger. Int J Heat Mass Transfer 45:1033–1044

Huang SC, Chang YP (1980) Heat conduction in unsteady, periodic, and steady states in laminated composites. J Heat Transfer 102(4):742–748

Ismail LS, Velraj R, Ranganayakulu C (2010) Studies on pumping power in terms of pressure drop and heat transfer characteristics of compact plate-fin heat exchangers—a review. Renew Sust Energ Rev 14(1):478–485

Jokar A, Eckels SJ, Hosni MH (2004) Thermo-hydrodynamic of the evaporation of refrigerant R134A in brazed plate heat exchangers. In: ASME heat transfer/fluids engineering summer conference, HT-FED2004-56573

Jonsson H, Moshfegh B (2000) Modeling of the thermal and hydraulic performance of plate fin, strip fin, and pin fin heat sinks – influence of flow bypass. In: Kromann GB, Culham JR, Ramakrishna K (eds) Intersociety conf on thermal and thermomechanical phenomena in electronic systems, vol 1, pp 185–192

Kameoka T, Nakamura K (1975) Investigation on convective heat transfer from finned plate surface, JSME, Transactions, vol. 41, no. 346, 1975–1976, p. 1878–1888.) Heat Transfer - Japanese Research, vol. 6, Jan.-Mar. 1977, p. 41–54. Translation

Kanzaka M, Iwabuchi M, Aoki Y, Ueda S (1989) Study on heat transfer characteristics of pin finned plate type heat exchangers. AIChE Symp Ser 269(85):306

Kays WM, London AL (1984) Compact hear exchangers, 3rd edn. McGraw-Hill, New Year

Kedzierski MA (1997) Effect of inclination on the performance of a compact brazed plate condenser and evaporator. Heat Transfer Eng 18(3):25–38

Kedzierski M (1998) Effect of inclination/mal-distribution on the performance of a compact brazed plate condenser and evaporator. ASHRAE Trans 104:1

Kern DQ, Kraus AD (1972) Extended surface heat transfer

Kim SY, Paek JW, Kang BH (2000) Flow and heat transfer correlations for porous fin in a plate-fin heat exchanger. Heat Transfer 122:572–578

Kim NH, Yun JH, Webb RL (1997) Heat transfer and friction correlations for wavy plate fin-and-tube heat exchangers. J Heat Transfer 119(3):560–567

Klett J, Ott R, McMillan A (2000) Heat exchangers for heavy vehicles utilizing high thermal conductivity graphite foams. SAE Paper No. 2000-01-2207

Koyama S, Yara T (1999) Heat transfer of binary Zeotropic mixtures in a plate-fin condenser. In: Proc int conf on compact heat exchangers and enhancement technology for the process ind, pp 423–430

Koyama S, Yu J, Matsumoto T (1998) Approximate analysis for laminar film condensation of pure refrigerant on vertical finned surface. J Enhanc Heat Transf 5(3):191–200

Kraus AD (1988) Sixty-five years of extended surface technology (1922–1987). Appl Mech Rev 41 (9):321–364

Laor K, Kalman H (1992) The effect of tip convection on the performance and optimum dimensions of cooling fins. Int Commun Heat Mass Transfer 19(4):569–584

Lee YN (1980) Technological advancement in all aluminum plate-type oil cooler. In: Compact heat exchangers – history, technological advancement and mechanical design problems HTD, ASME, vol 10, pp 145–152

Legkiy VM, Babenko Y, Dikiy VA (1979) Heat transfer and drag of plate-type heat exchangers with hemispherical projections. Heat Transfer Sov Res 11(2):143–150

Li Q, Flamant G, Yuan X et al (2011) Compact heat exchangers: a review and future applications for a new generation of high temperature solar receivers. Renew Sust Energ Rev 15 (9):4855–4875

Liang CY, Yang WJ (1975) Heat transfer and friction loss performance of perforated heat exchanger surfaces. J Heat Transfer 97:9–15

London AL, Shah RK (1968) Offset rectangular plate-fin surfaces heat transfer and flow friction characteristics. ASME J Eng Power 90:218–228

Longo GA, Gasparella A, Sartori R (2003) Development of innovative plate heat exchangers for refrigeration application. In: Proc of the 21st IIR int congress of refrigeration, Washington, DC, ICR 62

Longo GA, Gasparella A, Sartori R (2004a) Experimental heat transfer coefficients and pressure drop during refrigerant vaporization inside plate heat exchangers. In: Tenth international refrigeration and air conditioning conference at Purdue, R095

Longo GA, Gasparella A, Sartori R (2004b) Experimental heat transfer coefficients during refrigerant vaporization and condensation inside herringbone-type plate heat exchangers with enhanced surfaces. Int J Heat Mass Transfer 47:4125–4136

Manglik RM (1996) Plate heat exchangers for process industry applications: enhanced thermal hydraulic characteristics of chevron plates. In: Manglik RM, Kraus AD (eds) Enhanced and multiphase heat trans. Begell House, New York, pp 267–276

Marr YN (1990) Correlating data on heat transfer in plate-fin heat exchangers with short offset fins. Therm Eng 37:249–252

Marriott J (1977) Performance of an Alfaflex plate heat exchanger. Chem Eng Prog 2:73–78

Marsi MA, Cliffe KR (1996) A study of the deposition of fine particles in compact plate fin heat exchangers. J Enhanc Heat Transf 3:259–272

Marvillet C (1991) Welded plate heat exchangers as refrigerants dry-ex evaporators. Design and operation of heat exchangers. Springer-Verlag, Berlin, Germany, pp 265–268

Masri MA, Cliffe KR (1996) A study of the deposition of fine particles in compact plate fin heat exchangers. J Enhanc Heat Transf 3(4):259–272

McNab GS, Meisen A (1973) Thermophoresis in liquids. J Colloid Interface Sci 44(2):339–346

Mennicke U (1972) The apparent overall heat transfer coefficient of plate heat exchangers. Waerme Stoffuebertragung 5:168–180

Metzner AB, Friend WL (1958) Theoretical analogies between heat, mass and momentum transfer and modifications for fluids of high Prandtl or Schmidt numbers. Can J Chem Eng 36 (6):235–240

Mochizuki S, Yagi Y (1977) Heat transfer and friction characteristics of strip fins. Heat Transfer Jpn Res 6:36–59

Mochizuki S, Yagi Y, Enomoto T (1979) Transient response of air-cooled plate-fin condensers. Refrigeration 54(624):835–843

Molki M, Yuen CM (1986) Effect of interwall spacing on heat transfer and pressure drop in a corrugated wall channel. Int J Heat Mass Transfer 29:987–997

Muley A, Manglik RM (1997a) Enhanced heat transfer characteristics of single-phase flows in a plate heat exchanger with mixed Chevron plates. J Enhanc Heat Transf 4(3):187

Muley A, Manglik RM (1997b) Experimental study of turbulent flow heat transfer and pressure drop in a plate heat exchanger with chevron plates. In: Oosthuizen PH, Armaly BF, Chen TS, Pepper DW, Acharya S (eds) ASME proc of the 32nd national heat trans conf, HTD, vol 346, pp 69–76

Newson IH, Bott TR, Hussain CI (1983) Studies of magnetite deposition from a flowing suspension. Chem Eng Commun 20(5–6):335–353

O'Brien JE, Sparrow EM (1982) Corrugated-duct heat transfer, pressure drop and flow visualization. J Heat Transfer 104:410–416

Ohadi MM, Salemi M, Dessiatoun S, Singh A (1995) EHD-enhanced pool boiling of r-123 in a parallel plate configuration. In: Fletcher LS Aihara T (eds) Proc of the ASME/JSME thermal eng joint conference, vol 2, pp 225–232

Okada K, Ono M, Tomimura T, Okuma T, Konno H, Ohtani S (1972) Design and heat transfer characteristics of new plate type heat exchanger. Heat Trans Jpn Res 1(1):90–95

Oshima T, Iuchi S (1974) Calculation method for cooler condensers with continuous plate finned tubes. Heat Trans Jpn Res 3(3):1–5

Ouazia B (2001) Evaporation heat transfer and pressure drop of HFC-134a inside a plate heat exchanger. In: Aminemi NK, Toma O, Rudland R, Crain E (eds) Proc of the ASME process industries division paper, pp 115–124

Paffenbarger J (1990) General computer analysis of multistream, plate-fin heat exchangers in compact heat exchangers: a festschrift for London AL. Hemisphere Publishing Corporation, Washington, DC, pp 727–746

Panchal CB, Hillis DL, Thomas A (1983) Convective boiling of ammonia and Freon 22 in plate heat exchangers. ASME-JSME Therm Eng Joint Conf Proc 2:261–268

Panitsidis H, Greham RD, Westwater JW (1975) Boiling of liquids in a compact plate-fin heat exchanger. Int J Heat Mass Transfer 18:37–42

Parker KO, Coombs, MG (1980) New developments in compact plate-fin heat exchangers. In: Compact heat exchangers history technological advancement and mechanical design problems, vol 10, pp 171–179

Pauley WR, Eaton JK (1988) Experimental study of the development of longitudinal vortex pairs embedded in a turbulent boundary layer. AIAA J 26:816–823

Peng XF, Peterson GP, Wang BX (1996) Flow boiling of binary mixtures in microchanneled plates. Int J Heat Mass Transfer 39:1257–1264

Prasad BS (1991) The performance prediction of multistream plate-fin heat exchangers based on stacking pattern. Heat Transfer Eng 12(4):58–70

Pritchard AM, Clarke RH, de Block MX (1992) Fouling of small passages in compact heat exchangers. In: Fouling mechanisms, theoretical and practical aspects, Eurotherm seminar, vol 23, pp 47–56

Pucci PF, Howard CP, Piersall CH Jr (1967) The single blow transient testing technique for compact heat exchanger surfaces. J Eng Power 89:29–39

Quazia B (2001) Evaporation heat transfer and pressure drop of hfc-134a inside a plate heat exchanger. In: Proc of 2001 ASME int mech eng congress and exposition PID-6, pp 115–123

Raghavan VR, Murthy MS (1983) On the selection of fin profiles for OTEC plate-fin evaporators. Energy Convers Manag 23(4):193–199

Rao B, Das S (2004) Effect of flow distribution to the channels on the thermal performance of the multipass plate heat exchangers. Heat Transfer Eng 25(8):48–59

Robertson JM, Lovegrove PC (1980) Boiling heat transfer with Freon 11 in brazed-aluminum plate-fin heat exchangers. ASME paper no 80-HT-58

Rosenblad G, Kullendorf A (1975) Estimating heat transfer rates from mass transfer studies on plate heat exchanger surfaces. Wiirme Stoffubertrag 8:187–191

Rovazhyanskiy LL, Atroshchenko VI, Kedrov MS (1977) Coefficients of heat transfer for condensation of low pressure steam in plate condensers with slot-like channels in a grid pattern. Heat Trans Sov Res 9(2):28–31

Rupani A, Molki M, Ohadi M, Franca F (2002) Flow boiling of R-134a in a minichannel plate evaporator with augmented surface. Heat Trans Proc 12th Int Heat Trans Conf 4:279–284

Rupani AP, Molki M, Ohandi MM, Franca FHR (2003) Enhanced flow boiling of r-134a in a minichannel plate evaporator. J Enhanc Heat Transf 10:1–8

Sarma PK, Chary SP, Rao VD (1989) Condensation of vertical plate fins of variable cross section-limiting solutions. Can J Chem Eng 67:937–941

Shah RK (1975a) Perforated heat exchanger studies. Part 1: Flow phenomena, noise and vibration. ASME Paper No. 75-WA/HT-8, ASME, New York

Shah RK (1975b) Perforated heat exchanger studies. Part 2: Heat transfer and flow friction characteristics. ASME Paper 75-WA/HT-9, ASME, New York

Sieder EN, Tate GE (1936) Heat transfer and pressure drop of liquids in tubes. Ind Eng Chem 28(12):1429–1435

Smith MC (1972) Performance analysis and model experiments for louvered fin evaporator core development. SAE Paper No. 720078

Sohrabpour S, Razani A (1993) Optimization of convective fin with temperature-dependent thermal parameters. J Franklin Institute 330(1):37–49

Sparrow EM, Baliga RR, Patankar SV (1977) Heat transfer and fluid flow analysis of interrupted-wall channels, with application to heat exchangers. J Heat Transfer 99:4–11

Sparrow EM, Hossfeld M (1984) Effect of rounding protruding edges on heat transfer and pressure drop in a duct. Int J Heat Mass Transfer 27:1715–1723

Sparrow EM, Liu CH (1979) Heat transfer, pressure drop and performance relationships for inline, staggered, and continuous plate heat exchangers. Int J Heat Mass Transfer 22:1613–1625

Syed A (1991) The use of plate heat exchangers as evaporators and condensers in process refrigeration. In: Foumeny EA, Heggs PJ (eds) Heat trans eng design of heat exchangers. Ellis Horwood Limited, Hemel Hempstead, vol 1, no 10, pp 139–157

Taler D (2001) Mathematical model and experimental study of a plate-fin-and-tube heat exchanger (in Polish). Arch Automot Eng 4:145–162

Taler D (2004a) Determination of heat transfer correlations for plate-fin-and-tube heat exchangers. Heat Mass Transfer 40(10):809–822

Taler D (2004b) Experimental determination of heat transfer and friction correlations for plate fin-and-tube heat exchangers. J Enhanc Heat Transf 11(3):183–204

Talik AC, Fletcher LS, Anand NK, Swanson LW (1995a) Heat transfer and pressure drop characteristics of a plate heat exchanger using a propylene-glycol/water mixture as a working fluid. In: Sernas V, Boyd RB, Jensen MK (eds) Proc of the 30th national heat trans conf, HTD, vol 314, pp 83–88

Talik AC, Swanson LW, Fletcher LS, Anand NK (1995b) Heat transfer and pressure drop characteristics of a plate heat exchanger. In: Proc of the ASME/JSME thermal eng joint conference, vol 4, pp 321–330

Theoclitus G (1966) Heat transfer and flow-friction characteristics in nine pin-fin surfaces. J Heat Transfer 85:383–390

Thonon B, Grandgeorge S, Jallut C (1999) Effect of geometry and flow conditions on particulate fouling in plate heat exchangers. Heat Transfer Eng 20(3):12–24

Thonon B, Vidil R, Marvillet C (1995) Recent research and developments in plate heat exchangers. J Enhanc Heat Transf 2(1–2):149

Tiggelbeck S, Mitra NK, Fiebig M (1994) Comparison of wing-type vortex generators for heat transfer enhancement in channel flows. J Heat Transfer 116:880–885

Tishchenko ZV Bondarenko VN (1983) Comparison of the efficiency of smooth-finned plate heat exchangers. Int Chem Eng 23(3):550–557

Torii K, Nishina K, Nakayama K (1994) Mechanism of heat transfer augmentation by longitudinal vortices in a flat plate boundary layer. In: Heat transfer proc 10th int heat trans conf, vol 5, pp 123–128

Torikoshi K, Kawabata K (1989) Heat transfer and flow friction characteristics of a mesh finned air-cooled heat exchanger convection heat transfer and transport processes. In: Figliola RS, Kaviany M, Ebadian MA (eds) ASME symp, vol 116. ASME, New York, pp 71–77

Turner CW, Lister DH (1991) A study of the deposition of silt onto the surface of type 304 stainless steel. Can J Chem Eng 69(1):203–211

Vasil'ev VY (2006) An experimental investigation into rational enhancement of convective heat transfer in rectangular interrupted ducts of plate-fin heat-transfer surfaces. Therm Eng 53 (12):1006–1016

Vlasogiannis P, Karagiannis G, Argyropoulos P, Bontozoglou V (2002) Air-water two-phase flow and heat transfer in a plate heat exchanger. Int J Multiphase Flow 28:757–772

Wadekar VV, Robertson JM (1989) Two-phase pressure gradients in the vertical upflow boiling of cyclohexane in perforated plate-fin passages. AIChe Symp Ser 269(85):301–305

Wang CC, Lee CJ, Chang CT, Chang YJ (1999) Some aspects of plate fin- and-tube heat exchangers: with and without louvers. J Enhanc Heat Transfer 6(5):357

Wang Z, Li Y, Zhao M (2015) Experimental investigation on the thermal performance of multi-stream plate-fin heat exchanger based on genetic algorithm layer pattern design. Int J Heat Mass Transfer 82:510–520

Wang Z, Li Y (2016) Layer pattern thermal design and optimization for multistream plate-fin heat exchangers—a review. Renew Sust Energ Rev 53:500–514

Wang L, Sundén B (2001) Thermal performance analysis of multi-stream plate heat exchangers. Proc of the 35th National heat trans conf, ASME, New York, NY, paper no. NHTC2001–20052

Wang L, Sundén B (2003) Optimal design of plate heat exchangers with and without pressure drop specifications. Appl Therm Eng 23:295–311

Wang LK, Sunden B, Yang QS (1999) Pressure drop analysis of steam condensation in a plate heat exchanger. Heat Transfer Eng 20(1):71–77

Wanniarachchi AS, Ratman U, Tilton BE, Dutta-Roy K (1995) Approximate correlations for chevron-type plate heat exchangers. In: Sernas V, Boyd RB, Jensen MK (eds) Proc of the 30th national heat trans conf, HTD, vol 314, pp 145–152

Watel B, Thonon B (2001) An experimental study of convective boiling in a compact serrated plate-fin heat exchanger. J Enhanc Heat Transf 9(1):1–16

Webb RL (1987) In: Kakac S, Shah RK, Aung W (eds) Handbook of single-phase heat transfer, vol 17. John Wiley, New York, pp 1–17.62, Chapter 17

Webb RL (1994) The flow structure in the louvered fin heat exchanger geometry. In: Alkidas AC (ed) Vehicle thermal management. SAE, pp 221–232

Webb RL, Kim NY (2005) Principles of enhanced heat transfer. Taylor and Francis, New York

Webb RL (2018) Compact heat exchangers. J Enhanc Heat Transf 25(1):1–59

Wellsandt S, Vamling, L (2000) Heat transfer of R-22 and alternatives in plate- type evaporator. In: Tree DR (ed) Proceedings of the 2000 international refrigeration conference at Purdue, pp 1161–1168

Wong LT, Smith MC (1966) Airflow phenomena in the louvered-fin heat exchanger. SAE Paper No. 730237

Yan YY, Lin TF (1999) Evaporation heat transfer and pressure drop of refrigerant R-134a in a plate heat exchanger. J Heat Transfer 121(1):118–127

Yan YY, Lio HC, Lin TF (1999) Condensation heat transfer and pressure drop of refrigerant r-134a in a plate heat exchanger. Int J Heat Mass Transfer 42:993–1006

Yeh RH (1994) Optimum spines with temperature dependent thermal parameters. Int J Heat Mass Transfer 37(13):1877–1884

Yeh RH (1997) Analysis of thermally optimized fin array in boiling liquids. Int J Heat Mass Transfer 40(5):1035–1044

Yeh RH (2001) Optimum finned surfaces with longitudinal rectangular fins. J Enhanc Heat Transf 8 (4):279–289

Yeh RH, Liaw SP (1993) Optimum configuration of a fin for boiling heat transfer. J Franklin Institute 330(1):153–163

Zhang L, Qian Z, Deng J, Yin Y (2015) Fluid–structure interaction numerical simulation of thermal performance and mechanical property on plate-fins heat exchanger. Heat Mass Transfer 51 (9):1337–1353

Chapter 2
Offset-Strip Fins

Offset-strip fin OSF is by and large the most important heat transfer enhancement concept developed for gases. The OSF and louver fin geometries are used for both plate-and-fin and finned tube heat exchangers. In OSF, a laminar boundary layer is developed on the short strip length which is dissipated in the wake region between strips, Figs. 2.1 and 2.2. The j/f ratio is an efficiency index. The OSF yields 150% increase in thermal performance with j/f 0.83. Greater enhancement will be obtained by using shorter strip lengths.

DeJong and Jacobi (1997), Fig. 2.3, have worked with OSF and they have observed enhancement due to feathery and sinusoidal wake and large-scale vortex shedding when the wake closely resembled classical von Karman vortex street. With the increase in Reynolds number, the vortex shedding moves upstream and the entire array sheds vortices. At the low Reynolds number, however, the kinetic energy is not sufficient for the vortex shedding to occur and the Sherwood number decreases slightly. DeJong and Jacobi (1997) measured mass transfer coefficients using naphthalene sublimation test and they observed a peak in the Sherwood number with the increase in Reynolds number, coinciding with locations of the onset of vortex.

Shah and London (1978) predicted heat transfer coefficients for the thermal entry length laminar flow between parallel plates at constant temperature. However, their analysis was based on simplified approximation to full momentum and energy equations and they did not consider any complex flow characteristics like vortex shedding. The OSF and plain fin performance is obtained using VG-1. The approximate prediction model of j and f versus Reynolds number curves for the OSF has been given by Kays (1972). His model was based on the assumption that there is laminar boundary layer on the fins and the boundary layers developed on the fin are totally dissipated in the wake region between fins. The equations for laminar flow over a flat plate in a free stream have been used in the analysis. He has obtained

S. K. Saha et al., *Heat Transfer Enhancement in Plate and Fin Extended Surfaces*,
SpringerBriefs in Applied Sciences and Technology,
https://doi.org/10.1007/978-3-030-20736-6_2

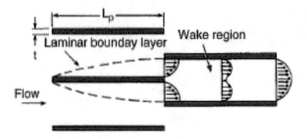

Fig. 2.1 Boundary layer and wake region of the OSF (Webb, 1987)

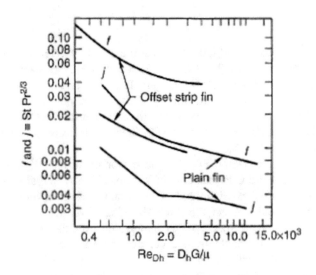

| | Surface Geometry | |
	Plain	Offset strip
Surface designation	10–27	10–58[a]
Fins/m	437	437
Plate spacing (mm)	12.2	11.9
Hydraulic diam (mm)	3.51	3.51
Fin thickness (mm)	0.20	0.24
Offset strip length in flow dir. (mm)	—	6.6

[a]Dimensions geometrically scaled to give same hydraulic diameter as plain fin.

Fig. 2.2 Comparison of j and/for the OSF and the plain-fin surface geometries (Webb, 1987)

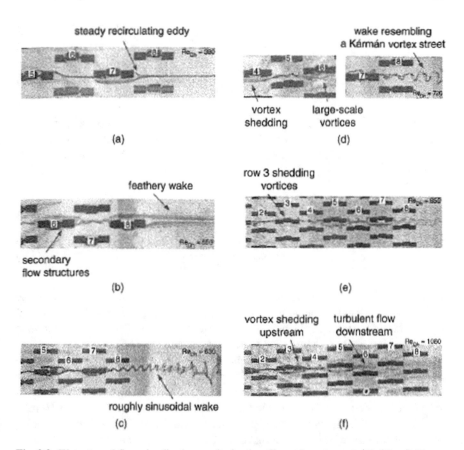

steady recirculating eddy

wake resembling
a Kármán vortex street

vortex   large-scale
shedding   vortices

(a)   (d)

feathery wake

row 3 shedding
vortices

secondary
flow structures

(b)   (e)

vortex shedding   turbulent flow
upstream   downstream

roughly sinusoidal wake

(c)   (f)

Fig. 2.3 Water tunnel flow visualization results for the offset-strip geometry with $LP = 2.54$ cm, $s = 1.28$ cm, $t = 0.32$ cm: (a) $Re_{Dh} = 380$, (b) $Re_{Dh} = 550$, (c) $Re_{Dh} = 630$, (d) $Re_{Dh} = 720$, (e) $Re_{Dh} = 850$, (f) $Re_{Dh} = 1060$ (DeJong and Jacobi 1997)

$$j = 0.664Re_L^{-0.5} \tag{2.1}$$

$$f = \frac{C_p t}{2L_p} + 1.328Re_L^{-0.5} \tag{2.2}$$

The first term in Eq. (2.2) above accounts for the form drag on the plate. Milne-Thomson (1960) have given $C_p = 0.88$ based on potential flow normal to a thin plate. The model neglects heat transfer and friction on the confining walls of the OSF channel. This approximation model, however, well serves the designer to predict the effect of strip length and thickness.

A more rational approach is to solve numerically the momentum and energy equations following the works of Patankar and his associates (Patankar 1990). Joshi and Webb (1987) gave an analytical model to predict the j and f versus Re_{Dh} characteristic curve of the OSF array. Their Nu and f equations are as follows:

$$\eta_{Nu} = \frac{1-\gamma}{(1+\alpha+\delta)^2} \left[(1+\delta)\eta_f Nu_p + \alpha(1+\alpha)Nu_e\right] \qquad (2.3)$$

$$f = \frac{1-\gamma}{(1+\gamma)(1+\alpha+\delta)} \left[f_p + \alpha f_e + \frac{C_p t}{2L_p}\right] \qquad (2.4)$$

Joshi and Webb (1987) model does a fairly good job. It is comparable to that of Manglik and Bergles (1990) and Muzychka and Yovanovich (2001a) correlations. Joshi and Webb (1987) determined the point of transition from laminar flow to turbulent flow as follows:

$$Re_{Dh,tr} = 257\left(\frac{L_p}{s}\right)^{1.23}\left(\frac{t}{L_p}\right)^{0.58}\left(\frac{D_h}{\delta_{mom}}\right) \qquad (2.5)$$

$$\text{where } \delta_m = t + 1.328\frac{L_p}{Re_L^{0.5}} \qquad (2.6)$$

is the fin thickness, plus twice the momentum thickness of the boundary layer at the end of the strip length. The laminar Nu_p and f_p terms in Eqs. (2.3) and (2.4) may be obtained from the numerical solution of Sparrow and Liu (1979) for $\alpha = 0$ and zero-thickness plates, which is modified to account for the effect of fin thickness.

Wieting (1975) and Joshi and Webb (1987) have developed correlations to predict j and f versus Re_{Dh} characteristics for the OSF array by multiple regression and power law expression. Manglik and Bergles (1990) used an asymptotic correlation method developed by Churchill and Usagi (1972) to provide another correlation for the OSF. However, the asymptotic values for small and high Reynolds number values must be known in this case. Usami (1991) is an important reference for OSF-enhanced surfaces.

Manglik and Bergles correlations are as follows:

$$f = 9.624 Re_{Dh}^{-0.742}\left(\frac{s}{h}\right)^{-0.186}\left(\frac{t}{L_p}\right)^{0.305}\left(\frac{t}{s}\right)^{-0.266}$$
$$\left[1 + 7.669 \times 10^{-8} Re_{Dh}^{4.43}\left(\frac{s}{h}\right)^{0.92}\left(\frac{t}{L_p}\right)^{3.77}\left(\frac{t}{s}\right)^{0.236}\right]^{0.1} \qquad (2.7)$$

$$j = 0.652 Re_{Dh}^{-0.540}\left(\frac{s}{h}\right)^{-0.154}\left(\frac{t}{L_p}\right)^{0.350}\left(\frac{t}{s}\right)^{-0.068}$$
$$\left[1 + 5.269 \times 10^{-5} Re_{Dh}^{1.34}\left(\frac{s}{h}\right)^{0.504}\left(\frac{t}{L_p}\right)^{0.456}\left(\frac{t}{s}\right)^{-1.06}\right]^{0.1} \qquad (2.8)$$

$$\text{where } D_h = \frac{4sbL_p}{\left[2\left(sL_p + bL_p + tb\right) + ts\right]} \qquad (2.9)$$

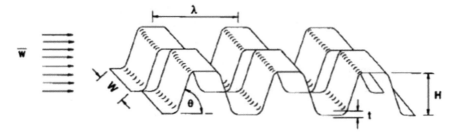

Fig. 2.4 Schematic diagram of turbulator geometry (Muzychka and Yovanovich 2001a)

Table 2.1 Summary of surface characteristics (Muzychka and Yovanovich 2001a)

Device	F_w	F_f	ϕ	S_o	d_h
CPI—1	1.916	0.4783	0.8769	2.225	2.231
CPI—2	1.921	0.4190	0.8769	2.225	2.484
CPI—3	2.383	0.5804	0.8953	1.815	1.822
CPI—4	2.080	0.5194	0.8869	1.731	2.068
CPI—5	2.417	0.5863	0.8869	1.850	1.780
SQ—1	2.049	0.6319	0.9167	3.223	2.729
SQ—2	1.984	0.6223	0.9089	3.111	2.556
SQ—3	1.899	0.6096	0.8963	2.968	2.312
SQ—4	2.457	0.5931	0.9091	2.361	2.067
SQ—5	2.441	0.5904	0.9091	2.361	2.081

The previously mentioned correlation for air (gas) ($Pr = 0.7$) was extended to liquids like water, polyalphaolefin ($3 < Pr < 150$), and automotive oils ($250 < Pr < 775$) by Hu and Herold (1995) and Muzychka and Yovanovich (2001b). Very few data are available on the OSF geometry for liquids and at high Prandtl number.

Per cent fin offset has an effect on the enhanced surface performance. 0.5 offset performance is different from that with 0.3 offset (Kurosaki et al. 1988; Hatada and Senshu 1984). The wake length varies from one strip width in the flow direction to the several strip widths. The existence of burrs on the upstream and downstream fin edges during the manufacture of the OSF geometry has a definite effect on the OSF performance (Webb and Joshi 1983).

Muzychka and Yovanovich (2001a) prepared analytic models for transverse flow through an offset-strip fin array. These models were developed for low-flow asymptotic behavior with laminar and turbulent boundary layer wake models. Thermal-hydraulic characteristic behavior was investigated by using fundamental equations of fluid dynamics and heat transfer. The results of this model were compared with new experimental data for ten offset-strip fin configurations. A turbulator strip is an offset-strip fin which is used in many automotive oil coolers. Figure 2.4 shows the geometry of turbulator.

Patankar et al. (1977) studied the heat transfer characteristics and pressure drop behavior for transfer flow through an array of interrupted plates and staggered plate by numerical methods. Model predictions were agreed with ±20% for 92% of

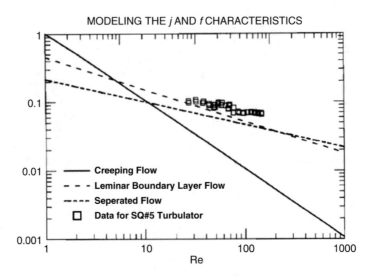

Fig. 2.5 Asymptotic limits of Colburn j factor (Muzychka and Yovanovich 2001a)

Table 2.2 Comparison of models with data using optimal values of blending parameters (Muzychka and Yovanovich 2001a)							
		f			j		
Device	m	n	RMS	p	q	RMS
CPI—1	1	0.99	5.80	3.0	4.0	10.21
CPI—2	–	–	–	–	–	–
CPI—3	1	0.85	4.23	4.0	5.0	13.42
CPI—4	1	1.00	3.95	5.0	5.0	17.69
CPI—5	1	0.87	4.85	5.0	5.0	29.50
SQ—1	1	0.79	3.59	5.0	1.25	12.31
SQ—2	1	0.71	2.98	5.0	1.25	7.82
SQ—3	1	0.80	1.49	5.0	1.75	9.97
SQ—4	1	0.85	2.11	4.5	5.0	35.92
SQ—5	1	0.81	5.27	5.0	4.0	9.37

friction factor data and 71% for Colburn j factor data. Table 2.1 shows the surface characteristics of turbulator. Figure 2.5 shows the asymptotic behavior of Colburn j factor with respect to Reynolds number. Table 2.2 presents comparison of models with data using optimal values of correlation parameters for combining the creeping flow, laminar boundary layer, and turbulent layer. Table 2.3 presents the root mean square (RMS) and (min/max) values for different models where fixed values of correlation parameter were used. Bergles (1985), Webb (1987, 1994), and Kalinin et al. (1998) reviewed on enhanced heat transfer by using different types of fins.

Muzychka and Yovanovich (2001b) presented thermal-hydraulic characteristics of offset-strip fin arrays of short ducts or channels. They developed different models by combining the asymptotic behavior for laminar and turbulent wake regions.

Table 2.3 Comparison of models with data with fixed values of blending parameter (Muzychka and Yovanovich 2001a)

Device	$f(m = 1, n = 6/7)$		$j\ (p = 9/2, q^a)$	
	RMS	(min/max)	RMS	(min/max)
CPI—1	8.25	−16.70/7.12	16.27	0.38/26.66
CPI—2	18.12	−37.50/20.08	17.44	−29.28/ −7.70
CPI—3	6.57	−2.70/10.30	13.37	−22.31/−3.38
CPI—4	16.89	−29.98/−10.16	17.72	−34.98/4.28
CPI—5	5.94	−6.36/12.27	29.53	−41.18/−10.27
SQ—1	11.50	6.40/18.05	12.06	−25.47/23.20
SQ—2	18.88	14.58/23.97	8.95	−19.90/18.26
SQ—3	9.44	7.00/12.46	17.99	−28.95/−0.40
SQ—4	5.84	2.59/11.25	35.90	−52.41/−17.40
SQ—5	10.76	−0.30/17.91	9.38	−21.13/12.97

[a] $q = 7/5$ for straight profiles, $q = 5$ for curved or rounded profiles

Table 2.4 Definitions of hydraulic diameter for offset-strip fin models (Muzychka and Yovanovich 2001b)

Model	Definition of d_h	
Wieting (1975)	$(2sH/(s + H))$	(i)
Joshi and Webb (1987)	$(2(s - t)HL_f/(sL_f + HL_f + tH))$	(ii)
Manglik and Bergles (1990)	$(4sHL_f/(2(sL_f + HL_f + tH) + ts))$	(iii)

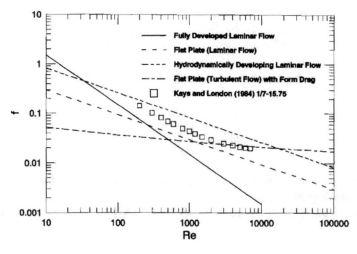

Fig. 2.6 Comparison of asymptotic solutions with data (Kays and London 1984)

These models were compared with experimental data for 19 configurations of the rectangular offset-strip fin. The result obtained from these models had within ±20% for 96% of friction factor and 82% for Colburn j factor data. Table 2.4 shows the formulas to determine the hydraulic diameter for offset-strip fin models. Friction factor is flow characteristic behavior of duct at low Reynolds number. Figure 2.6 shows the comparison of asymptotic solutions with data obtained from Kays and London (1984).

Offset-strip fin (OSF) is an array of short noncircular ducts, and friction factor is an important factor for flow characteristic behavior in a duct at low Reynolds number. Webb and Joshi (1982) developed a simple model by using the Churchill and Usagi (1972) correlation method which was valid only for OSF having aspect ratio less than 0.25. Muzychka and Yovanovich (2001b) presented a model to overcome the limitation of aspect ratio which was made by Webb and Joshi (1982). If limit $Re_{dh} \rightarrow 0$, the value of f is tending such that flow becomes fully developed in a rectangular channel and if limit $Re_{dh} \rightarrow \infty$ the value of f becomes a constant which represented the drag component due to the finite fin thickness. Shapiro et al. (1954) gave relation between apparent friction factor and Reynolds number in the hydrodynamic entrance region of straight duct:

$$f_{app} = \frac{3.44}{\sqrt{Z^+}} \tag{2.10}$$

where $Z^+ = L_f/D_h Re_{Dh}$ is the dimensionless sub-channel length, d_h = hydraulic diameter of the OSF array, D_h = hydraulic diameter of the rectangular sub-channel.

Muzychka (1999) predicted the correlation for flat plate and fully developed flow in laminar region:

$$f_{lam} = \frac{f Re_{D_h}(d_h/D_h)}{Re_{d_h}} + 1.328 \left(Re_{d_h} \frac{L_f}{d_h} \right)^{-\frac{1}{2}} \tag{2.11}$$

Kays and London (1984) predict the turbulent friction factor for a typical OSF array:

$$f_{tur} = 0.074 \left(Re_{d_h} \frac{L_f}{d_h} \right)^{-\frac{1}{5}} + \frac{H_t + \frac{St}{2}}{2L_f(H+S)} C_D \tag{2.12}$$

where H = Channel height, S = Width of sub-channel, t = Thickness of fin, C_D = Form drag coefficient.

Churchill and Usagi (1972) correlation is as follows for laminar-transition-turbulent region:

$$f = \left[\left\{ \frac{f Re_{D_h}(d_h/D_h)}{Re_{d_h}} + 1.328 \left(Re_{d_h} \frac{L_f}{d_h} \right)^{-\frac{1}{2}} \right\}^n + \left\{ 0.074 \left(Re_{d_h} \frac{L_f}{d_h} \right)^{-\frac{1}{5}} + \frac{Ht + St/2}{2L_f(H+S)} C_D \right\}^n \right]^{\frac{1}{n}} \tag{2.13}$$

where "n" is the correlation parameter.

The value of Colburn j factor approaches fully developed in the limit of $Re_{dh} \rightarrow 0$ and at limit $Re_{dh} \rightarrow \infty$ its value approaches for boundary layer flow. Kays and Crawford (1993) gave a correlation for OSF in the laminar duct flow region in terms of Reynolds number:

$$j = 0.664(Re_{L_f})^{-\frac{1}{2}} \tag{2.14}$$

where $L_f = $ fin length.

London and Shah (1968) proposed a generalized solution for thermally developing flow:

$$j = 0.641 \left[\frac{fRe_{D_h} D_h}{Re_{D_h}^2 L_f} \right]^{\frac{1}{3}} \tag{2.15}$$

Muzychka and Yovanovich (1998a, b) developed correlations for j and f as below:

In the laminar region,

$$j_{lam} = \left[\left(\frac{Nu_{D_h}(d_h/D_h)}{Re_{d_h} Pr^{1/3}} \right)^5 + \left(\frac{0.641 fRe_{D_h}}{Re_{d_h}^{7/3}} \left(\frac{d_h^2}{D_h L_f} \right)^{1/3} \right)^5 \right]^{\frac{1}{5}} \tag{2.16}$$

Colburn j factor for turbulent region:

$$j_{tur} = 0.037 \left(Re_{d_h} \frac{L_f}{d_h} \right)^{-\frac{1}{5}} \tag{2.17}$$

Churchill and Usagi (1972) correlation method for laminar-transition-turbulent is as follows:

$$j = \left[\left\{ \left(\frac{Nu_{D_h}(d_h/D_h)}{Re_{d_h} Pr^{1/3}} \right)^5 + \left(\frac{0.641 fRe_{D_h}}{Re_{d_h}^{7/3}} \left(\frac{d_h^2}{D_h L_f} \right)^{1/3} \right)^5 \right\}^{m/5} + \left\{ 0.037 \left(Re_{d_h} \frac{L_f}{d_h} \right)^{-1/5} \right\}^m \right]^{1/m} \tag{2.19}$$

Song et al. (2017) conducted the numerical simulation for correlation of heat transfer and flow friction characteristics of offset-strip fins on the basis of optimization of heat exchangers. He studied the thermal-hydrodynamic characteristics by using fluent. They simulated the model under different values of Reynolds numbers, then correlated, and verified these results with the Manglik and Bergles correlations. The schematic of the off-strip fins has been shown in Fig. 2.7. Different heat transfer and friction factor correlations for off-strip fins have been presented in Table 2.5.

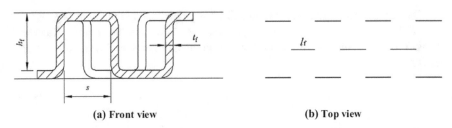

(a) Front view   (b) Top view

Fig. 2.7 Schematic of offset-strip fins (Song et al. 2017)

Shah and Bhatti (1987), Najafi et al. (2011), Sanaye and Hajabdollahi (2010), Kays and London (1984), London and Shah (1968), Wieting (1975), Joshi and Webb (1987), Mochizuki et al. (1987), Kurosaki et al. (1988), Dubrovsky and Vasiliev (1988), Ismail and Velraj (2009), Wen et al. (2012), and Manglik and Bergles (1995) carried out a series of experiments and numerical simulation on developing the correlation of the flow friction and heat transfer characteristics of the different types of fins.

Ferrouillat et al. (2006), Ismail et al. (2009), Lu et al. (2011), Kim et al. (2011), and Xu et al. (2015) adopted standard k-ε model, RNG k-ε model, and laminar model to investigate the thermo-flow characteristics under different ranges of Reynolds number. Table 2.6 presents different correlations for heat transfer and flow characteristics of the offset-strip fins. Table 2.7 presents the geometry parameter of fin1. Table 2.8 shows the specification of the common offset-strip fins and Table 2.9 shows the structure parameters. Numerical simulation results for friction factor and heat transfer with respect to Reynolds number and Manglik and Bergles calculated result correlation are presented in Figs. 2.8, 2.9, and 2.10. The average relative deviation between numerical calculated results is summarized in Table 2.10.

Bhowmik and Lee (2009) worked on 3D numerical modelling for heat transfer and pressure drop characteristic of an offset-strip fin heat exchanger. The offset-strip fin has properties like decreased boundary layer thickness, high heat transfer coefficients, and higher heat transfer area per unit volume. Kays and London (1984), Manson (1950), Wieting (1975), Joshi and Webb (1987), Mochizuki et al. (1987), Manglik and Bergles (1995), Tinaut et al. (1992), and Hu and Herold (1995) worked on establishing the correlations for offset-strip fins and plate-fin heat exchangers.

They modelled offset-strip fin module as presented in Fig. 2.11 assumed incompressible steady fluid and constant thermophysical properties irrespective of the variation of temperature. They plotted f and j-factor, Fig. 2.12, versus Reynolds number and validated results which are within 10–20%. Due to the fact that heat transfer coefficient depends on channel aspect ratio, heat transfer mechanism, and geometric parameters, some deviation was found as geometries are different in literature and this study. Bhowmik and Lee (2009) established correlation

$$f_{cor} = 10 Re_{dh}^{-0.68} \tag{2.20}$$

$$j_{cor} = 0.489 Re_{dh}^{-0.445} \tag{2.21}$$

Table 2.5 Different heat transfer and friction factor correlations for offset-strip fins (Song et al. 2017)

Index	Expressions	Notes
Wieting correlation	$Re_h \leq 1000$, $j = 0.483(l_f/D_h)^{-0.162}\xi^{-0.184}Re_h^{-0.536}$ $f = 7.661(l_f/D_h)^{-0.384}\xi^{-0.092}Re_h^{-0.712}$ $Re_h \geq 2000$, $j = 0.242(l_f/D_h)^{-0.322}(t_f/D_h)^{0.089}Re_h^{-0.368}$ $f = 1.136(l_f/D_h)^{-0.781}(t_f/D_h)^{0.534}Re_h^{-0.198}$	$D_h = \dfrac{2sh_f}{s + h_f}$ $\xi = s/h_f$ $Re_h = \dfrac{\rho u D_h}{\mu}$
Joshi and Webb correlation	$Re_h \leq Re^*$, $j = 0.53(l_f/D_h)^{-0.15}\xi^{-0.14}Re_h^{-0.5}$ $f = 8.12(l_f/D_h)^{-0.41}\xi^{-0.02}Re_h^{-0.74}$ $Re_h \geq Re^* + 1000$, $j = 0.21(l_f/D_h)^{-0.24}(t_f/D_h)^{0.02}Re_h^{-0.4}$ $f = 1.12(l_f/D_h)^{-0.65}(t_f/D_h)^{0.17}Re_h^{-0.36}$	$Re^* = 257\left(\dfrac{l_f}{s}\right)^{1.23}\left(\dfrac{t_f}{l_f}\right)^{0.58} D_h$ $\left[t_f + 1.328\dfrac{l_f}{Re_l^{0.5}}\right]^{-1}$ $D_h = \dfrac{2(s - t_f)h_f}{(s + h_f) + \dfrac{t_f h_f}{l_f}}$ $Re_l = \dfrac{\rho u l_f}{\mu}$
Mochizuki correlation	$Re_h < 2000$, $j = 1.37(l_f/D_h)^{-0.25}\xi^{-0.184}Re_h^{-0.67}$ $f = 5.55(l_f/D_h)^{-0.32}\xi^{-0.092}Re_h^{-0.67}$ $Re_h \geq 2000$, $j = 1.17\left(\dfrac{l_f}{D_h} + 3.75\right)^{-1}(t_f/D_h)^{0.089}Re_h^{-0.36}$ $f = 0.83\left(\dfrac{l_f}{D_h} + 0.33\right)^{-0.5}(t_f/D_h)^{0.534}Re_h^{-0.20}$	$D_h = \dfrac{2sh_f}{s + h_f}$ $\xi = s/h_f$ $Re_h = \dfrac{\rho u D_h}{\mu}$

(continued)

Table 2.5 (continued)

Index	Expressions	Notes
Manglik and Bergles correleation	$j = 0.6522 Re_h^{-0.5403} \xi^{-0.1541} \delta^{0.1499} \eta^{-0.0678}$ $\times \{1 + 5.269 \times 10^{-5} Re_h^{1.340} \xi^{0.504} \delta^{0.456} \eta^{-1.055}\}^{0.1}$ $f = 5.55(l_f/D_h)^{-0.32} \xi^{-0.092} Re_h^{-0.67}$ $\times \{1 + 1.76669 \times 10^{-8} Re_h^{4.429} \xi^{0.92} \delta^{3.767} \eta^{0.236}\}^{0.1}$	$\xi = s/h_f, \quad \delta = t_f/l_f, \quad \eta = t_f/s$ $D_h = \dfrac{4sh_fl_f}{2(sl_f + h_fl_f + t_fh_f) + t_fs}$
ALEX correlation	$300 \le Re_h \le 7500,$ $\ln j = -2.64136 \times 10^{-2}(\ln Re_h)^3 + 0.555843(\ln Re_h)^2$ $\quad - 4.09241(\ln Re_h) + 6.21681$ $\ln f = 0.132856(\ln Re_h) - 2.28042(\ln Re_h) + 6.79634$	$D_h = \dfrac{2sh_f}{s + h_f}$ $\xi = s/h_f$ $Re_h = \dfrac{\rho u D_h}{\mu}$

Where l_f is the offset length of fins, m; t_f is the thickness, m; s is the fin spacing, m; h_f is the fin height, m

Table 2.6 Correlation for heat transfer and flow friction characteristics of the offset-strip fins (Song et al. 2017)

Index	Expressions	Notes
Wieting correlation [6]	$Re_h \leq 1000$, $j = 0.483(l_f/D_h)^{-0.162}\xi^{-0.184}Re_h^{-0.536}$ $f = 7.661(l_f/D_h)^{-0.384}\xi^{-0.092}Re_h^{-0.712}$ $Re_h \geq 2000$, $j = 0.242(l_f/D_h)^{-0.322}(t_f/D_h)^{0.089}Re_h^{-0.368}$ $f = 1.136(l_f/D_h)^{-0.781}(t_f/D_h)^{0.534}Re_h^{-0.198}$	$D_h = \dfrac{2sh_f}{s+h_f}$ $\xi = s/h_f$ $Re_h = \dfrac{\rho u D_h}{\mu}$
Joshi and Webb correlation [7]	$Re_h \leq Re^*$, $j = 0.53(l_f/D_h)^{-0.15}\xi^{-0.14}Re_h^{-0.5}$ $f = 8.12(l_f/D_h)^{-0.41}\xi^{-0.02}Re_h^{-0.74}$ $Re_h \geq Re^* + 1000$, $j = 0.21(l_f/D_h)^{-0.24}(t_f/D_h)^{0.02}Re_h^{-0.4}$ $f = 1.12(l_f/D_h)^{-0.65}(t_f/D_h)^{0.17}Re_h^{-0.36}$	$Re^* = 257\left(\dfrac{l_f}{s}\right)^{1.23}\left(\dfrac{t_f}{l_f}\right)^{0.58}$ $D_h\left[t_f + 1.328\dfrac{l_f}{Re_l^{0.5}}\right]^{-1}$ $D_h = \dfrac{2(s-t_f)h_f}{(s+h_f)+\dfrac{t_f h_f}{l_f}}$ $Re_l = \dfrac{\rho u l_f}{\mu}$
Mochizuki correlation [8]	$Re_h < 2000$, $j = 1.37(l_f/D_h)^{-0.25}\xi^{-0.184}Re_h^{-0.67}$ $f = 5.55(l_f/D_h)^{-0.32}\xi^{-0.092}Re_h^{-0.67}$ $Re_h \geq 2000$, $j = 1.17\left(\dfrac{l_f}{D_h}+3.75\right)^{-1}(t_f/D_h)^{0.089}Re_h^{-0.36}$ $f = 0.83\left(\dfrac{l_f}{D_h}+0.33\right)^{-0.5}(t_f/D_h)^{0.534}Re_h^{-0.20}$	$D_h = \dfrac{2sh_f}{s+h_f}$ $\xi = s/h_f$ $Re_h = \dfrac{\rho u D_h}{\mu}$

(continued)

Table 2.6 (continued)

Index	Expressions	Notes
Manglik and Bergles correleation [11]	$j = 0.6522Re_h^{-0.5403}\xi^{-0.1541}\delta^{0.1499}\eta^{-0.0678}$ $\times\{1 + 5.269 \times 10^{-5}Re_h^{1.340}\xi^{0.504}\delta^{0.456}\eta^{-1.055}\}^{0.1}$ $f = 5.55(l_f/D_h)^{-0.32}\xi^{-0.092}Re_h^{-0.67}$ $\times\{1 + 1.76669 \times 10^{-8}Re_h^{4.429}\xi^{0.92}\delta^{3.767}\eta^{0.236}\}^{0.1}$	$\xi = s/h_f,\ \ \delta = t_f/l_f,\ \ \eta = t_f/s$ $D_h = \dfrac{4sh_fl_f}{2(sl_f + h_fl_f + t_fh_f) + t_fs}$
ALEX correlation [26]	$300 \le Re_h \le 7500,$ $\ln j = -2.64136 \times 10^{-2}(\ln Re_h)^3 + 0.555843(\ln Re_h)^2$ $-4.09241(\ln Re_h) + 6.21681$ $\ln f = 0.132856(\ln Re_h) - 2.28042(\ln Re_h) + 6.79634$	$D_h = \dfrac{2sh_f}{s + h_f}$ $\xi = s/h_f$ $Re_h = \dfrac{\rho u D_h}{\mu}$

Where l_f is the offset length of fins, m; t_f is the thickness, m; s is the fin spacing, m; h_f is the fin height, m

Table 2.7 Summary of geometry parameters of fin1 (Song et al. 2017)

Parameter	h_f/mm	s/mm	t_f/mm	l_f/mm
fin1	9.3	1.8	0.2	5

Table 2.8 Specification of the common Chinese offset-strip fins (Song et al. 2017)

Parameter	h_f/mm	s/mm	t_f/mm	l_f/mm
fin1	9.3	1.8	0.2	5
fin2	9.3	12	0.2	5
fin3	9.3	1.5	0.2	5
fin4	6.2	1.4	0.3	5
fin5	6.2	1.7	0.3	5
fin6	4.4	1.7	0.3	5
fin7	2.7	3.2	0.3	5

Table 2.9 Structure parameters of the common Chinese offset-strip fins (Song et al. 2017)

Parameter	ξ	δ	η
fin1	0.193	0.040	0.111
fin2	0.129	0.040	0.167
fin3	0.161	0.040	0.133
fin4	0.226	0.060	0.214
fin5	0.274	0.060	0.176
fin6	0.386	0.060	0.176
fin7	1.185	0.060	0.094

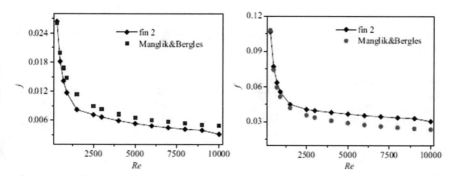

Fig. 2.8 Numerical simulation results and calculation results of Manglik and Bergles of fin2 (Song et al. 2017)

and studied the effect of Prandtl number using these correlations as presented in Fig. 2.13. They evaluated the performance of compact heat exchanger on three parameters, j/f, $j/f^{1/3}$, and thermohydraulic performance JF (Fig. 2.14), and concluded that for $Pr = 7$ and $Pr = 50$, the convenient performance criteria were JF and $j/f^{1/3}$, respectively.

Hong and Cheng (2009) presented three-dimensional numerical analysis for microelectronic cooling by using water in offset-strip fin microchannel heat sinks.

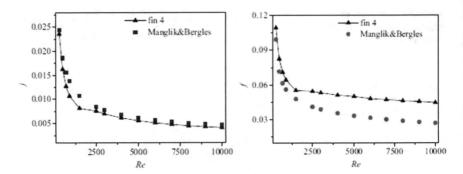

Fig. 2.9 Numerical simulation results and calculation results of Manglik and Bergles of fin4 (Song et al. 2017)

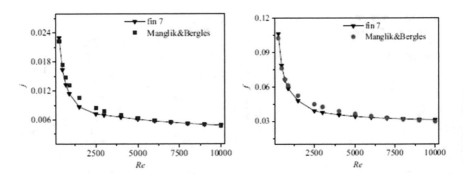

Fig. 2.10 Numerical simulation results and calculation results of Manglik and Bergles of fin7 (Song et al. 2017)

Table 2.10 Average relative deviations between the numerical and calculated results (Song et al. 2017)

Parameter	j	f
fin2	19.56%	19.26%
fin4	12.58%	38.00%
fin7	6.48%	5.35%

Kandlikar and Upadhye (2005) and Colgan et al. (2007) worked on strip-fin microchannels. Peles et al. (2005), Koşar and Peles (2006), Koşar et al. (2005), and Siu-Ho et al. (2007) investigated micro-pin fin for heat transfer enhancement.

Hong and Cheng (2009) considered base layer as inlet and outlet occupancy in microchannel heat sink was small and typically made up of low thermal conductive materials which have no impact on heat transfer. This study focuses on fin interval and fin length presuming that channel width (W_c), channel depth (H_c), and wall thickness (W_w) are fixed. They defined K as the ratio of fin performance to fin length and M as the number of fins so that performance can be evaluated on the basis of K and M values. They assumed that (1) Navier-Stokes equation and non-slip wall boundary conditions are admissible; (2) flow is incompressible, laminar, and steady

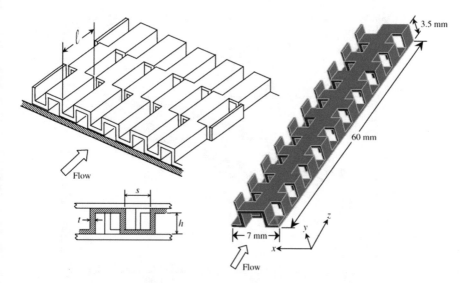

Fig. 2.11 Geometry of the offset-strip fin modules (Bhowmik and Lee 2009)

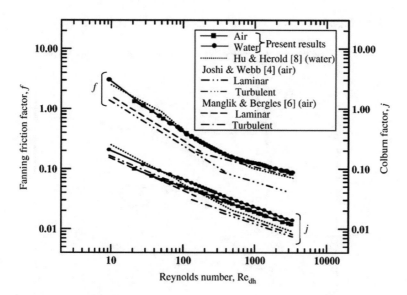

Fig. 2.12 Variation in the Fanning factor and Colburn factor with Reynolds number (Bhowmik and Lee 2009)

state, (3) viscous dissipation and gravity effect on fluid flow is negligible, and (4) thermophysical properties of water and silicon wafer are temperature dependent for numerical formulation. They numerically investigated velocity, vorticity, and temperature fields, Fig. 2.15.

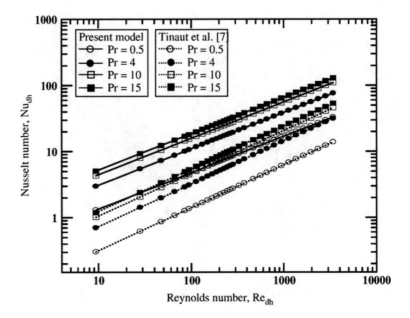

Fig. 2.13 Variation in Nu_{dh} with Re_{dh} for different Pr (Bhowmik and Lee 2009)

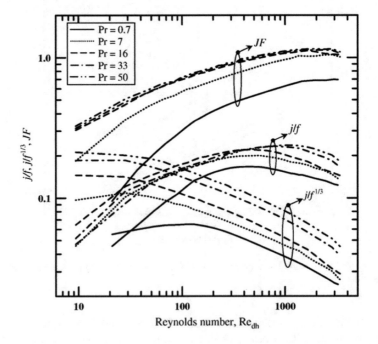

Fig. 2.14 Variation in performance factors with Re_{dh} (Bhowmik and Lee 2009)

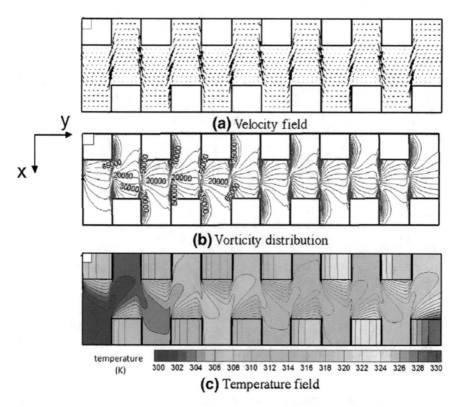

Fig. 2.15 Fluid flow and heat transfer characteristics at the middle plane of a strip-fin microchannel with $K = 1$ and $M = 6$. (**a**) The velocity, (**b**) vorticity, (**c**) temperature distribution (x:y:z = 1:30:3 and the flow direction is from the left to the right) (Hong and Cheng 2009)

The periodical blockage generates large vorticity indicating large circulation of flow. This provides mixing of colder and heated coolant enhancing heat transfer due to periodic breakage of boundary layer. They accounted for the effect of K and M and presented the results in Fig. 2.16. From Fig. 2.16a they concluded that mass flow decreases as M increases for fixed K. Also, Fig. 2.16a depicts that for a fixed M, mass flow rate increased convective surface area. Figure 2.16b shows agreement between lower mass flow rate and higher flow resistance directing to highest value of M at which corresponding pressure drop is the lowest. Figure 2.16c indicates pumping power versus M graph and can be correlated from Fig. 2.16a and b as pumping power is the product of mass flow rate and pressure drop. Figure 2.16d shows the variation of temperature difference with K and M. The heat flux effect and maximum wall temperature behavior on microchannel heat sink have been presented in Fig. 2.17. They claimed that $K = 1$ was the optimal condition for offset-strip fin microchannel heat sink.

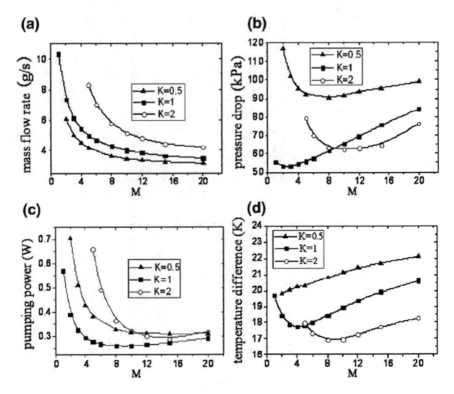

Fig. 2.16 Performance of offset-strip fin microchannel heat sinks with different K and M values for $q = 200$ W/cm^2 and $T_{s,max} = 332$ K (Hong and Cheng 2009)

Ismail et al. (2009) carried out numerical analysis of the heat transfer performance of compact heat exchanger in turbulent flow regime. They focused on the maldistribution of the fluid flow in the header and used baffles in order to achieve flow uniformity. They studied the flow maldistribution in three compact heat exchanger models. The thermohydraulic performance of the heat exchanger using 3 models of offset-strip fins and 16 designs of wavy fins has also been studied numerically. Their work resulted in obtaining a large set of design data for compact heat exchangers. The schematic of offset-strip fin and wavy fin has been shown in Fig. 2.18a and b, respectively. The header with a baffle and the baffle configuration have been shown in Fig. 2.19a and b, respectively.

Shah and Sekulic (2003), Ranganayakulu et al. (1997), Zhang et al. (2003), Lalot et al. (1999), Wen et al. (2006), Ranganayakulu et al. (1997, 2006), and Ranganayakulu and Seetharamu (1999) have studied the flow nonuniformity or flow maldistribution in compact heat exchangers and have proposed different header configurations. Mochizuki and Yagi (1975), Manson (1950), and Maiti (2002) studied thermohydraulic performance of offset-strip fins.

Fig. 2.17 Performance of offset-strip fin microchannel heat sinks with different K and M values: (**a**) $q = 100$ W/cm^2 and $T_{s,max} = 332$ K and (**b**) $q = 100$ W/cm^2 and $T_{s,max} = 323.5$ K (Hong and Cheng 2009)

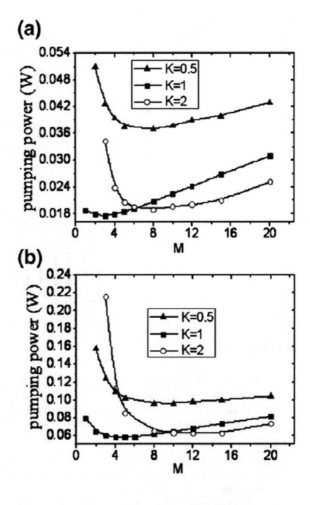

Fig. 2.18 PFHE with serrated fins (Wang et al. 2009)

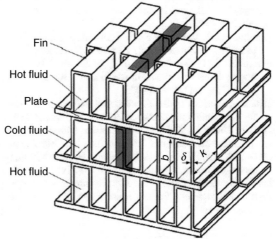

Fig. 2.19 Schematic of
offset-strip fin and wavy fin
(Sheik Ismail et al. 2009)

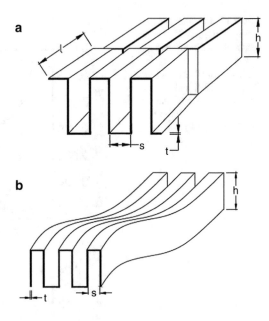

References

Bergles AE (1985) Chapter 3–Techniques to augment heat transfer. In: Handbook of heat transfer applications, 2nd edn. McGraw-Hill, New York

Bhowmik H, Lee KS (2009) Analysis of heat transfer and pressure drop characteristics in an offset strip fin heat exchanger. Int Commun Heat Mass Transfer 36(3):259–263

Churchill SW, Usagi R (1972) A general expression for the correlation of rates of transfer and other phenomena. AICHE J 18(6):1121–1128

Colgan EG, Furman B, Gaynes M, Graham WS, LaBianca NC, Magerlein JH, Marston KC (2007) A practical implementation of silicon microchannel coolers for high power chips. IEEE Trans Compon Packag Technol 30(2):218–225

DeJong NC, Jacobi AM (1997) An experimental study of flow and heat transfer in parallel-plate arrays: local, row-by-row and surface average behaviour. Int J Heat Mass Transfer 40:1365–1378

Dubrovsky EV, Vasiliev VY (1988) Enhancement of convective heat transfer in rectangular ducts of interrupted surfaces. Int J Heat Mass Transfer 31(4):807–818

Ferrouillat S, Tochon P, Garnier C, Peerhossaini H (2006) Intensification of heat transfer and mixing in multifunctional heat exchangers by artificially generated streamwise vorticity. Appl Therm Eng 26(16):1820–1829

Hatada T, Senshu T (1984) Experimental study on heat transfer characteristics of convex louver fins for air conditioning heat exchangers. ASME paper 84-HT-74, New York

Hong F, Cheng P (2009) Three dimensional numerical analyses and optimization of offset strip-fin microchannel heat sinks. Int Coummun Heat Mass Transfer 36(7):651–656

Hu S, Herold KE (1995) Prandtl number effect on offset fin heat exchanger performance: predictive model for heat transfer and pressure drop. Int J Heat Mass Transfer 38(6):1043–1051

Ismail LS, Velraj R (2009) Studies on Fanning friction (f) and Colburn (j) factors of offset and wavy fins compact plate fin heat exchanger-a CFD approach. Numer Heat Trans Part A Appl 56 (12):987–1005

Ismail LS, Ranganayakulu C, Shah RK (2009) Numerical study of flow patterns of compact plate-fin heat exchangers and generation of design data for offset and wavy fins. Int J Heat Mass Transfer 52(17):3972–3983

Joshi HM, Webb RL (1987) Heat transfer and friction in the offset strip fin heat exchanger. Int J Heat Mass Transfer 30(1):69–84

Kalinin EK, Dreitser GA, Kopp IZ, Myakochin AS (1998) Effektivnye poverkhnosti teploobmena (Effective Heat Transfer Surfaces), Moscow: Energoatomizdat

Kandlikar SG, Upadhye HR (2005) Extending the heat flux limit with enhanced microchannels in direct single-phase cooling of computer chips. In: Semiconductor thermal measurement and management IEEE twenty first annual IEEE symposium IEEE, pp 8–15

Kays WM (1972) Compact heat exchangers, AGARD lecture series on heat exchangers. 57 JJ Ginoux Ed, AGARD-LS-57-72

Kays WM, Crawford ME (1993) Convective heat and mass transfer. McGraw-Hill, New York

Kays WM, London AL (1984) Compact heat exchangers. McGraw-Hill, New York

Kim MS, Lee J, Yook SJ, Lee KS (2011) Correlations and optimization of a heat exchanger with offset-strip fins. Int J Heat Mass Transfer 54(9):2073–2079

Koşar A, Peles Y (2006) Thermal-hydraulic performance of MEMS-based pin fin heat sink. J Heat Transfer 128(2):121–131

Koşar A, Mishra C, Peles Y (2005) Laminar flow across a bank of low aspect ratio micro pin fins. J Fluid Eng 127(3):419–430

Kurosaki Y, Kashiwagi T, Kobayashi H, Uzuhashi H, Tang SC (1988) Experimental study on heat transfer from parallel louvered fins by laser holographic interferometry. Exp Therm Fluid Sci 1 (1):59–67

Lalot S, Florent P, Lang SK, Bergles AE (1999) Flow maldistribution in heat exchangers. Int J Appl Therm Eng 19:847–863

London AL, Shah RK (1968) Offset rectangular plate-fin surfaces—heat transfer and flow friction characteristics. J Eng Power 90(3):218–228

Lu CW, Huang JM, Nien WC, Wang CC (2011) A numerical investigation of the geometric effects on the performance of plate finned-tube heat exchanger. Energy Convers Manag 52 (3):1638–1643

Manglik RM, Bergles AE (1990) The thermal-hydraulic design of the rectangular offset strip-fin compact heat exchanger. In: Shah RK, Kraus AD, Metzger D (eds) Compact heat exchangers. Hemisphere, Washington, DC, pp 123–150

Manglik RM, Bergles AE (1995) Heat transfer and pressure drop correlations for the rectangular offset strip fin compact heat exchanger. Exp Therm Fluid Sci 10(2):171–180

Manson SV (1950) Correlations of heat transfer data and of friction data for interrupted plane fins staggered in successive rows, NACA technical note, page no. 2234–2237

Maiti DK (2002) Heat transfer and flow friction characteristics of plate-fin heat exchanger surfaces – a numerical study, PhD thesis, IIT Kharagpur, India

Milne-Thomson LM (1960) Theoretical hydrodynamics, 4th edn. Macmillan, New York, p 319

Mochizuki S, Yagi S (1975) Heat transfer and friction characteristics of strip fins. Int J Refrig 50:36–59

Mochizuki S, Yagi Y, Yang WJ (1987) Transport phenomena in stacks of interrupted parallel-plate surfaces. Int J Exper Heat Transfer 1(2):127–140

Muzychka YS, Yovanovich M (1998a) Modeling friction factors in non-circular ducts for developing laminar flow. In: Second AIAA theoretical fluid mechanics meeting, p 2492

Muzychka YS, Yovanovich M (1998b) Modeling Nusselt numbers for thermally developing laminar flow in non-circular ducts. In: Seventh AIAA/ASME joint thermophysics and heat transfer conference, p 2586

Muzychka YS (1999) Analytical and experimental study of fluid friction and heat transfer in low Reynolds number flow heat exchangers, PhD thesis, University of Waterloo, Waterloo, ON

Muzychka YS, Yovanovich MM (2001a) Modeling the f and j characteristics for transverse flow through an offset strip fin at low Reynolds number. J Enhanc Heat Transf 8(4):243–259

Muzychka YS, Yovanovich MM (2001b) Modeling the f and j characteristics of the offset strip fin array. J Enhanc Heat Transf 8(4):261–277

Najafi H, Najafi B, Hoseinpoori P (2011) Energy and cost optimization of a plate and fin heat exchanger using genetic algorithm. Appl Therm Eng 31(10):1839–1847

Patankar SV (1990) Numerical prediction of flow and heat transfer in compact heat exchanger passages. In: Shah RK, Kraus AD, Metzger D (eds) Compact heat exchangers. Hemisphere, Washington, DC, pp 191–204

Patankar SV, Liu CH, Sparrow EM (1977) Fully developed flow and heat transfer in ducts having streamwise-periodic variations of cross-sectional area. J Heat Transfer 99(2):180–186

Peles Y, Koşar A, Mishra C, Kuo CJ, Schneider B (2005) Forced convective heat transfer across a pin fin micro heat sink. Int J Heat Mass Transfer 48(17):3615–3627

Ranganayakulu C, Seetharamu KN (1999) The combined effects of longitudinal heat conduction, flow nonuniformity and temperature nonuniformity in crossflow plate-fin heat exchangers. Int J Commun Heat Mass Transfer 26:669–678

Ranganayakulu C, Seetharamu KN, Sreevatsan KV (1997) The effects of inlet fluid flow nonuniformity on thermal performance and pressure drops in crossflow plate-fin heat exchangers. Int J Heat Mass Transfer 40(1):27–38

Ranganayakulu Ch, Sheik Ismail L, Vengudupathi C (2006) Uncertainties inestimation of Colburn (j) factor and fanning friction (f) factor for offset stripfin and wavy fin compact heat exchanger surfaces. In: Mishra SC, Prasad BVSSS, Garimella SV (eds) Proceedings of the XVIII national and VII ISHMT –ASME heat and mass transfer conference, Guwahati, India, pp 1096–1103

Sanaye S, Hajabdollahi H (2010) Thermal-economic multi-objective optimization of plate fin heat exchanger using genetic algorithm. Appl Energy 87(6):1893–1902

Shah RK, Bhatti MS (1987) Handbook of single-phase convective heat transfer. Wiley, New York

Shah RK, London AL (1978) Laminar flow forced convection in ducts, supplement 1 to advances in heat transfer. Academic Press, New York

Shah RK, Sekulic DP (2003) Fundamentals of heat exchanger design. John Wiley & Sons, New York

Shapiro AH, Siegel R, Kline SJ (1954) Friction factor in the Lamianar entry region of a smooth tube. Proceedings of the 2nd U.S. National Congress of applied mechanics, pp 733–741

Siu-Ho A, Qu W, Pfefferkorn F (2007) Experimental study of pressure drop and heat transfer in a single-phase micropin-fin heat sink. J Electron Packag 129(4):479–487

Song R, Cui M, Liu J (2017) A correlation for heat transfer and flow friction characteristics of the offset strip fin heat exchanger. Int J Heat Mass Transfer 115:695–705

Sparrow EM, Liu CH (1979) Heat transfer, pressure drop and performance relationships for inline, staggered, and continuous plate heat exchangers. Int J Heat Mass Transfer 22:1613–1625

Tinaut FV, Melgar A, Ali AR (1992) Correlations for heat transfer and flow friction characteristics of compact plate-type heat exchangers. Int J Heat Mass Transfer 35(7):1659–1665

Usami H (1991) Pressure drop characteristics of OSF surfaces. In: Lloyd JR, Kurosake Y (eds) Proc 1991 ASME IJSME joint thermal engineering conference, vol 4. ASME, New York, pp 425–432

Wang YQ, Dong QW, Liu MS, Wang D (2009) Numerical study on plate-fin heat exchangers with plain fins and serrated fins at low Reynolds number. Chem Eng Technol 32(8):1219–1226

Webb RL (1987) Chapter 17–Enhancement of single-phase heat transfer. In: Kakac S, Shah RK, Aung W (eds) Handbook of single-phase heat transfer. Wiley, New York, pp 17.1–17.62

Webb RL, Joshi HM (1982) A friction factor correlation for the offset strip-fin matrix. Heat Trans 1982, vol 6. Hemisphere Publishing Company, pp 257–262

Webb RL, Joshi HM (1983) Prediction of the friction factor for the offset strip-fin matrix. In: ASME-JSME thermal eng joint conf, vol 1. ASME, New York, pp 461–470

Webb RL (1994) The flow structure in the louvered fin heat exchanger geometry. In: Alkidas AC (ed) Vehicle thermal management. SAE, pp 221–232

Wen J, Li YM, Wang SM, Li YZ, Wu CL (2012) Fluid flow and heat transfer characteristics in plain fins of plate-fin heat exchanger. Chem Eng (China) 40(10):25–28. (in Chinese)

Wen J, Yanzhong L, Zhou A, Zhang K (2006) An experimental and numerical investigation of flow patterns in the entrance of plate-fin heat exchanger. Int J Heat Mass Transf 49:1667–1678

Wieting AR (1975) Empirical correlations for heat transfer and flow friction characteristics of rectangular offset-fin plate-fin heat exchangers. J Heat Transfer 97(3):488–490

Xu S, Zhu HM, Yu R (2015) Numerical study on plate-fin heat exchangers with plain fin. Comput Appl Chem 32(8):977–981. (in Chinese)

Zhang J, Muley A, Borghess JB, Manglik RM (2003) Computational and experimental study of enhanced laminar flow heat transfer in three dimensional sinusoidal wavy-plate-fin channels, Proc of the 2003 ASME summer heat trans conf, Nevada, USA, HT2003–47148

Chapter 3
Louver Fins and Convex Louver Fins

3.1 Louver Fins

Louver fin geometry has a similarity to OSF. The entire slit may be rotated 20–45° rather than offsetting the slit strips. This angle is relative to the airflow direction. The louver surface fin is the standard geometry for air-cooled automotive heat exchangers. Louver fin geometries have a louver strip width of 0.9–1.5 mm in the airflow direction. Louver fin geometry gives heat transfer coefficients comparable to that of OSF for the same strip width. Shah and Webb (1982) give detail information about louver fin geometries along with their modifications. Davenport (1983a, b) varied the louver dimensions systematically for two louver heights. Aoki et al. (1989) and Fujikake et al. (1983) studied the effect of louver angle and louver pitch.

Kajino and Hiramatsu (1987) and Achaichia and Cowell (1988) numerically studied the laminar flow in a variety of louver arrays. Sunden and Svantesson (1990) experimented with louver orientation angle not normal to the airflow. They observed that j factor is lower than that for the standard 90° orientation. Suga and Aoki (1991) performed a numerical investigation to determine the optimum value of the ratio of fin pitch to louver pitch for fixed louver angle. They obtained the optimum design balances between high heat transfer performance and pressure drop. Chang and Wang (1996) made a good investigation for louver fin.

Davenport experimented with the louver fin geometries given in Fig. 3.1 and their correlations are.

$$j = 0.249 Re_L{}^{-0.42} L_h{}^{0.33} H_{0.26} \left(\frac{L_L}{H}\right)^{1.1} \qquad 300 < Re_{Dh} < 4000 \qquad (3.1)$$

© The Author(s), under exclusive license to Springer Nature Switzerland AG 2020
S. K. Saha et al., *Heat Transfer Enhancement in Plate and Fin Extended Surfaces*,
SpringerBriefs in Applied Sciences and Technology,
https://doi.org/10.1007/978-3-030-20736-6_3

Fig. 3.1 Definition of
geometric parameters of the
louver-fin heat exchanger.
(**a**) Tube cross section, (**b**)
tube-fin array, (**c**) end view
of fins, (**d**) cross section of
louver region (Chang and
Wang 1997)

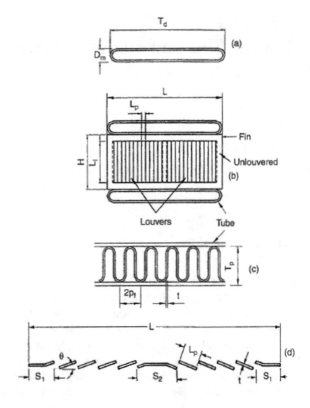

$$4f = 5.47Re_\text{L}^{-0.72}L_\text{h}^{0.37}L_\text{p}^{0.2}H^{0.23}\left(\frac{L_\text{L}}{H}\right)^{0.89} \qquad 70 < Re_\text{Dh} < 1000 \qquad (3.2)$$

$$4f = 0.494Re_\text{L}^{-0.39}H^{0.46}\left(\frac{L_\text{h}}{L_\text{p}}\right)^{0.33}\left(\frac{L_\text{p}}{H}\right)^{1.1} \qquad 1000 < Re_\text{Dh} < 4000 \qquad (3.3)$$

Equations (3.1) and (3.3) predict the heat transfer performance of automotive
radiators. However, Davenport correlation does not fully account for all dimensional
variables in the louver fin geometry. Dillen and Webb (1994) have developed a
correlation, based on the analytical model of Sahnoun and Webb (1992).

Plate-fin-and-tube louver fin geometry is included in the database of 91 samples
of louver fin heat exchangers (Davenport 1983b; Tanaka et al. 1984; Achaichia and
Cowell 1988; Webb 1988; Sunden and Svantesson 1990; Webb and Jung 1992;
Rugh et al. 1992; Chang and Wang 1996). Based on this database, correlation for
friction factor has been developed by Chang et al. (2000) and that for heat transfer
has been developed by Chang and Wang (1997). The correlations are given below:

$$j = Re_L^{-0.49} \left(\frac{\theta}{90} \right)^{0.27} \left(\frac{P_f}{L_p} \right)^{-0.14} \left(\frac{H}{L_p} \right)^{-0.29} \left(\frac{T_d}{L_p} \right)^{-0.23} \left(\frac{L_L}{L_p} \right)^{0.68} \left(\frac{T_p}{L_p} \right)^{-0.28} \left(\frac{t}{L_p} \right)^{-0.05}$$

(3.4)

$$f = f1 * f2 * f3$$

(3.5)

$$f1 = 14.39 Re_L^{-0.805 \frac{P_f}{H}} \left[\ln \left(1.0 + \left(\frac{P_f}{L_p} \right) \right) \right]^{3.04} \qquad Re_L \leq 150$$

(3.6)

$$f2 = \left[\ln \left(\frac{t}{P_f} \right)^{0.48} + 0.9 \right]^{-1.435} \left(\frac{D_h}{L_p} \right)^{-3.01} [\ln (0.5 Re_L)]^{-3.01} \qquad 150 < Re_L < 5000$$

(3.7)

$$f3 = \left(\frac{P_f}{L_L} \right)^{-0.308} \left(\frac{L}{L_L} \right)^{-0.308} \left(e^{-0.117 \frac{T_p}{D_m}} \right) \theta^{0.35} \qquad Re_L \leq 150$$

(3.8)

$$f3 = \left(\frac{T_p}{D_m} \right)^{-0.0446} \ln \left(1.2 + \left(\frac{L_p}{P_f} \right)^{1.4} \right)^{-3.553} \theta^{-0.447} \qquad 150 < Re_L < 5000$$

(3.9)

This correlation is applicable only to plate-fin geometries having inclined tube arrangements and not for staggered tube arrangements.

Beecher and Fagan (1987), Critoph et al. (1996, 1999), Eckels and Rabas (1987), Ha et al. (1998), Huang et al. (2003), Kayansayan (1993), Kim and Song (2003), Kim et al. (1997), Kundu and Das (1997), Kushida et al. (1986), Leu et al. (2004), Maltson et al. (1989), McQuiston (1978), Nacerbey et al. (2003), Rich (1975), Saboya and Sparrow (1974, 1976), Saboya et al. (1984), Seshimo and Fujii (1987, 1991), Taler (2004), Torikoshi et al. (1994), Tutar and Akkoca (2004), Wang and Chang (1998), Wang et al. (1996, 1999), and Yamashita et al. (1987) are the important works on plate-fin and tube type of heat exchangers.

The flow phenomena and performance characteristics of the louvered fin need to be further studied. Previously it was generally thought to be parallel to the louvers. But Davenport (1980) made flow visualization studies and he observed that the main flow stream did not pass through the louvers at low Re_L. At high Re_L when inertia force is strong, the flow becomes nearly parallel to the louvers. At low air velocities, the developing boundary layers on adjacent louvers become thick enough to effectively block the passage and this results in nearly axial flow through the array.

Figure 3.2 shows heat transfer characteristics of several louvered plate-fin geometries. At the highest Re_L, the data are parallel, but lower than those for laminar boundary flow over a flat plate (the flat line). The data represents the characteristics of laminar duct flow at low Re_L. Heat transfer performance depends on the flow structure and two types of flow here are:

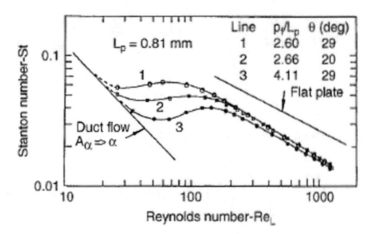

Fig. 3.2 Heat transfer characteristics of several louvered plate-fin geometries tested (Achaichia and Cowell 1988)

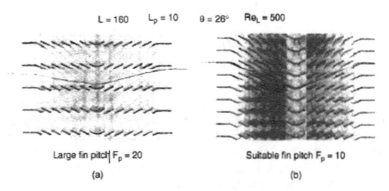

Fig. 3.3 Steaklines at $Re = 500$ in 160 mm deep flow visualization model of louver fin array ($I = 10$ mm, $8 = 26°$) (Kajino and Hiramatsu 1987). (**a**) $p1 = 20$ mm, (**b**) $p1 = 10$ mm

(a) Duct flow where the fluid travels axially through the array
(b) Boundary layer flow where the fluid travels parallel to the louvers

Figure 3.3, Kajino and Hiramatsu 1987, shows that for higher fin pitch, but same louver pitch, the significant fraction of the flow bypasses since the hydraulic resistance of the duct flow region is much less than that for boundary layer flow across the louvers. With lower fin pitch, again with same louver pitch, the hydraulic resistance of the duct increases and most of the flow passes through the louvers. Webb and Trauger (1991) defined flow efficiency η as

$$\eta = \frac{\text{Actual transverse distance}}{\text{Ideal transverse distance}} \qquad (3.10)$$

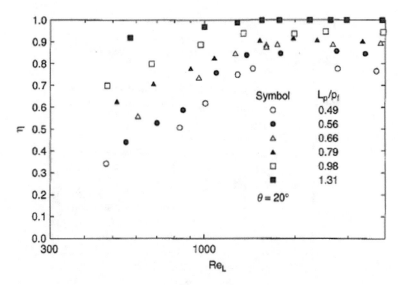

Fig. 3.4 Measured flow efficiency vs. Reynolds number for 20° louver angle (Webb and Trauger 1991)

They have made detailed flow visualization studies to simulate the flow in louver fin arrays. Sahnoun and Webb (1992) say

$$\eta = 0.95 \left(\frac{L_p}{P_f}\right)^{0.23} \qquad Re_L > Re_L^* \qquad (3.11)$$

$$\eta = \eta^* - 37.17 \times 10^{-6} (Re_L^* - Re_L)^{1.1} \left(\frac{L_p}{P_f}\right)^{-1.35} \left(\frac{\theta}{90}\right)^{-0.61} \qquad Re_L < Re_L^* \qquad (3.12)$$

where $\eta = \eta^*$ at $Re_L = Re_L^*$ and Re_L^*

$$= 828 \left(\frac{\theta}{90}\right)^{-0.34} \quad \text{with } \theta \text{ in degrees.} \qquad (3.13)$$

Figures 3.4 and 3.5 show measured flow efficiency and that predicted by various correlations, respectively. Some other flow efficiency predictive correlations are given by Bellows (1996), Achaichia and Cowell (1988), and Zhang and Tafti (2003), which include or exclude fin thickness.

Aoki et al. (1989), Fig. 3.6, measured heat transfer coefficient on individual louver fins using the two-dimensional array. The theoretical Pohlhausen solution, from Kays and Crawford (1980), is given by

$$\frac{h_{av}L}{k} = 0.906 Re_L^{1/2} Pr^{1/3} \qquad (3.14)$$

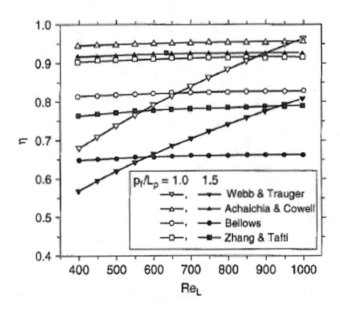

Fig. 3.5 The flow efficiency predicted by various correlations. Predictions were made at $P_f = 1.0$ and 1.5 mm, $L_P = 1.0$ mm, $t = 0.1$ mm, and $\theta = 30°$ (Webb and Kim 2005)

DeJong and Jacobi (2003) investigated vortex shedding in the louver geometry. Shedding from last downstream louver traced backwards in the upstream direction with the increase in Reynolds number. Vortex shedding does not produce as much enhancement for louver fins as that in case of OSF geometries. Since the distance between two consecutive louvers along a streamline is long, Dejong and Jacobi (2003) observed that the critical Reynolds number, when the vortex shedding starts, decreases as the louver angle increases and as $\frac{P_t}{L_p}$ decreases.

Suga and Aoki (1991) and Lyman et al. (2002) studied the effect of the distance between louvers in the flow direction on thermal wakes; thermal wake dissipation is more with longer wake length. The enhanced surface design must consider the optimum distance between louvers in the flow direction. The numerical study by Suga and Aoki (1991) revealed that the optimum design is given by

$$\frac{P_f}{L_p} = 1.5 \tan \theta \qquad (3.15)$$

This gives good balance between high heat transfer performance and pressure drop.

Flow in the louver fin is more complex than that in OSF geometry and Sahnoun and Webb (1992) analytically modelled j and f predictions like which were given by Joshi and Webb (1987) for OSF geometry. The heat transfer coefficient in the unlouvered end regions is predicted by using a fully developed laminar low

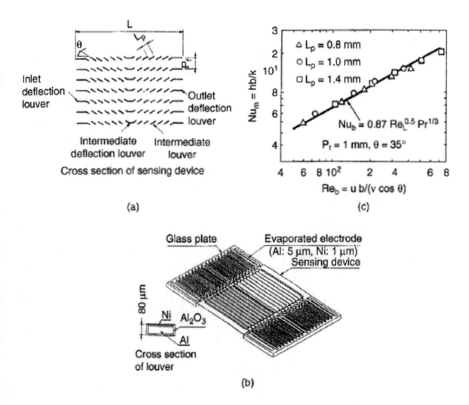

Fig. 3.6 Louver fin tests. (**a**) Louvered heat transfer plate, (**b**) array of louvered plates tested, (**c**) average experimental Nusselt numbers (Aoki et al. 1989)

situation. Similarly, the friction model works. The friction model takes care of pressure drag caused by the finite fin thickness. Davenport (1983b) compared the prediction.

The bypass effect was studied by Dillen and Webb (1994) by modifying Sahnoun and Webb (1992) model. Webb et al. (1995) extended the model development work. In all these models, the bypass effect was modelled using the flow efficiency equation.

Muzychka and Kenway (2009) investigated for predicting the thermohydraulic characteristic of offset-strip fin array for heat transfer enhancement. Muzychka and Yovanovich (2001), Michna et al. (2007), and Shah and Sekulic (2003) studied strip fins. They developed simple analytic model which accurately and precisely predicts the friction factor of f and Colburn factor j characteristic of the array for both laminar wake and turbulent wake regions. They experimentally collected eight strip-fin configuration data for SAE 5W30. They used counterflow heat exchanger cell and presented all relevant data in Table 3.1. The proposed modified new model is

Table 3.1 Summary of model and data comparisons (Muzychka and Kenway 2009)

Device	Medium	s, mm	W, mm	H, mm	t, mm	d_h, mm	Pr	Re	f, %rms	j, %rms
osf7	5W–30 oil	1.5	3.05	3.15	0.254	1.840	387–618	2–39	34.57	18.54
osf9	5W–30 oil	1.856	3.05	3.15	0.254	2.111	373–595	3–41	12.43	10.93
8292	5W–30 oil	1.4	3.4	3.15	0.254	1.769	404–623	2–33	15.61	28.00
90P	5W–30 oil	1.6	4.94	2.972	0.1016	2.019	402–618	2–39	10.99	23.77
*87P	5W–30 oil	2.2	3.89	2.972	0.1016	2.439	408–634	3–41	25.31	69.52
*93P	5W–30 oil	0.907	2.8	2.56	0.0508	1.311	380–589	2–25	13.85	55.01
*osf10	5W–30 oil	2.2	3.43	2.86	0.15	2.346	402–618	3–49	14.32	55.07
osf12	5W–30 oil	2.096	3.55	2.972	0.254	2.233	365–597	4–49	9.55	28.41
k11101935	Air	1.21	2.54	1.905	0.1	1.403	0.7	200–3000	16.31	13.63
k11101974	Air	1.24	2.54	1.29	0.05	1.221	0.7	200–3000	9.87	11.29
k11102703	Air	0.839	2.54	6.35	0.1	1.426	0.7	120–4000	9.10	23.43
k1121194D	Air	1.976	12.7	2.937	0.152	2.289	0.7	300–9000	11.84	4.57
k114154D	Air	1.499	6.35	2.54	0.152	1.805	0.7	300–6000	20.65	11.13
k1162118D	Air	1.983	4.52	4.41	0.102	2.664	0.7	200–9000	15.03	9.08
k1181612 T	Air	1.372	3.175	2.557	0.185	1.665	0.7	300–5000	3.51	10.02
k1171575D	Air	1.501	3.628	3.785	0.102	2.082	0.7	200–7000	16.88	16.80
k1181612D	Air	1.425	3.175	2.54	0.152	1.715	0.7	300–1500	8.75	7.03
k1181561	Air	1.525	3.17	6.35	0.1	2.283	0.7	300–6000	8.15	19.42
k1181600D	Air	1.435	3.175	3.162	0.152	1.866	0.7	300–5000	4.81	24.41
k1181982D	Air	1.181	3.175	2.527	0.102	1.514	0.7	200–4000	19.38	22.87
k1181986	Air	1.170	3.17	2.489	0.1	1.529	0.7	300–5000	3.90	16.72
k11182006D	Air	1.165	3.175	2.477	0.102	1.521	0.7	300–4000	3.62	19.85
k1192268	Air	1.020	3.17	7.65	0.1	1.745	0.7	120–5000	3.05	9.56
k1192412	Air	0.950	2.82	1.905	0.1	1.208	0.7	150–3000	9.15	12.27
k1192501	Air	0.916	2.82	5.08	0.1	1.498	0.7	120–4000	10.19	20.12
k1332122	Air	1.983	2.38	12.31	0.1	3.283	0.7	500–10.000	16.78	14.76

k1501MOD	Air	0.745	1.27	0.665	0.0254	0.678	0.7	50–800	14.81	11.19
Plate: 2	Water	1.65	3.18	2.34	0.152	1.810	3–8	121–2410	20.87	19.62
Plate: 3	Water	1.52	6.12	2.26	0.152	1.731	3–8	127–2500	23.09	12.33
Plate: 4	Water	1.28	3.33	3.84	0.152	1.825	3–8	73–1580	24.09	7.75
Plate: 5	Water	1.26	3.4	2.36	0.152	1.545	3–8	106–2230	21.33	5.60
Plate: 6	Water	1.55	3.33	2.367	0.152	1.757	3–8	124–2450	20.20	23.87
Plate: 7	Water	1.62	3.22	3.84	0.152	2.162	3–8	75–1280	20.74	17.15
Plate: 2	PAO	1.65	3.18	2.34	0.152	1.810	40–140	44–1000	3.62	2.15
Plate: 3	PAO	1.52	6.12	2.26	0.152	1.731	40–140	12–991	0.83	0.81
Plate: 4	PAO	1.28	3.33	3.84	0.152	1.825	40–140	9–561	1.56	0.22
Plate: 5	PAO	1.26	3.4	2.36	0.152	1.545	40–140	12–745	2.13	1.74
Plate: 6	PAO	1.55	3.33	2.367	0.152	1.757	40–140	12–910	1.79	1.45
Plate: 7	PAO	1.62	3.22	3.84	0.152	2.162	40–140	9–674	0.61	2.47

$$j = \left[\left\{ \left(\frac{NuD_hd_h}{ReD_hPr^{1/3}} \right)^5 + \left(\frac{0.641fRe_{dh}^{1/3}}{Re^{7/3}} \left(\frac{d_h^2}{D_hX} \right)^{1/3} \right)^5 \right\}^{7/10} + \left\{ 0.037 \left(Re \; d_h \frac{X}{d_h} \right)^{-1/5} \right\}^{7/2} \right]^{2/7}$$

$$(3.16)$$

They calculated 39 sets of data for SAE 5W30 oil, water, and PAO, respectively. They observed root mean square error between 0.61% and 34.57% whereas in the case of j factor this error reduces to 1–25%. They concluded that j factor ensures good prediction for higher Prandtl number fluids.

Suga and Aoki (1995) performed numerical simulation of multilouvered fins with a two-dimensional finite different code by using an overlaid grid method on heat transfer characteristics. Numerical results showed correlation between the fin geometries and heat transfer characteristics. They observed that multilouvered fins could be optimized by controlling thermal wakes after louvers. It was also viewed that heat exchanger performed superiorly at smaller louver angle.

Suga et al. (1990) developed new method by using overlaid grid method to solve the problem against generation of computational grids for multilouvered fin with a finite fin thickness. Figure 3.7 shows the corrugated multilouvered fins used in industrial applications. Toyoshima et al. (1986), Kajino and Hiramatsu (1987), and Tomoda and Suzuki (1988) had proposed the numerical simulation on the heat exchanger characteristics of multilouvered thin fins. Tanaka et al. (1984) and Suzuki et al. (1985) studied on inclined louvered fins and an offset-strip-fin-type heat exchanger, respectively. Figure 3.8 shows values of the mean Nusselt number and pressure drop as a function of Re_g for $Re = 450$, $\theta = 26°$, $P_f/P_t = 0.8$, and $t/P_t = 0.08$.

Elyyan and Tafti (2009) examined the dimpled multilouvered fin configuration and studied their heat transfer and flow characteristics. Webb and Trauger (1991), Tafti (1999), Tafti and Zhang (2001), Zhang and Tafti (2001), Lyman et al. (2002),

Fig. 3.7 Schematic view of corrugated multilouvered fins (Suga and Aoki 1995)

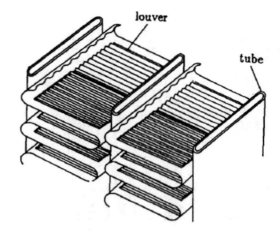

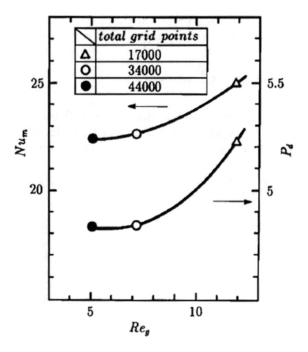

Fig. 3.8 The effect of grid refinement on the computed results (Suga and Aoki 1995)

and Dejong and Jacobi (2003) worked on louvered fin for heat transfer enhancement. Mahmood et al. (2001), Ligrani et al. (2001), Chyu et al. (1997), Moon et al. (1999), Burgess and Ligrani (2004), and Ekkad and Nasir (2003) studied the dimpled surface heat transfer characteristic.

The objective of this study was to take favor of the surface roughness, interrupted surfaces, and smaller scale discontinuities due to perforation and combine them to make novel dimpled louver fin-based fin geometry. For this, they studied three louver geometries. In case 1, larger dimple imprint diameter has been developed whereas in case 2 no such explicit arrangement was done although both case 1 and case 2 have identical geometries. In case 3, there was an opening or a perforation in the dimple. The dimple diameter effects were examined in case 1 and case 2 whereas in case 3 perforation consequences on heat transfer and flow characteristics were analyzed.

The assumptions made were constant heat flux boundary conditions, fully developed hydrodynamic and thermal conditions, and no slip boundary conditions at the fin surface. They used incompressible time-dependent Navier-Stokes and energy equations for solving this problem. They concluded that dimple imprint diameter had a nominal effect on both the heat transfer and flow. However in case 3, perforation or opening in the dimples significantly influences the flow and heat transfer. It primarily affected the dimple side of fin. As the flow was redirected towards protruded side, it was found that recirculation region around the dimple was vanished. It was observed that boundary layer regenerated from the edge of the perforation and enhances heat transfer in that region. The protrusion side favors the mixing of redirected ejecting

fluid flow. Finally, they calculated 12–50% increase in heat transfer coefficient due
to perforation although there was 60% increase in friction.

3.2 Convex Louver Fins

Convex louver fins are a variant of the OSF and they are known as offset convex
louver fin (OCLF), as shown in Fig. 3.9 (Hatada and Senshu 1984). In Fig. 3.10, type
I is OSF and type II is OCLF. Figure 3.11 shows the effect of the angle θ on the j and

Fig. 3.9 Fin geometries
(Hatada and Senshu 1984)

No.	θ (deg)	% Offset
1	0	0
2	12.8	23
3	17.4	33
4	24.6	53
5	9.7	20
6	17.4	33
7	20.7	42
8	24.6	53

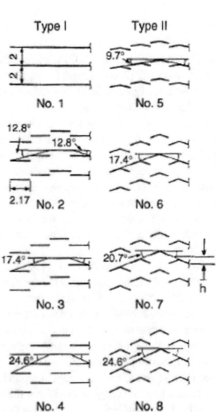

Fig. 3.10 j and f vs. Re_D for th18e louver geometries tested by Hatada and Senshu (1984). (**a**) OSF (type I), (**b**) OCLF (type II)

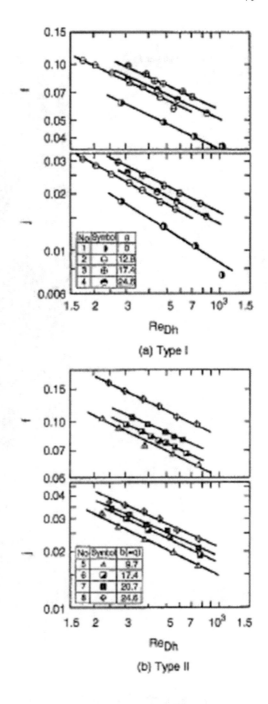

Fig. 3.11 Streamline flow
patterns in finned arrays
without tubes (Hatada and
Senshu 1984)

No powder region   Powder trace line

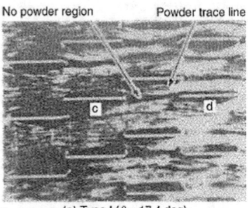

(a) Type I ($\theta = 17.4$ deg)

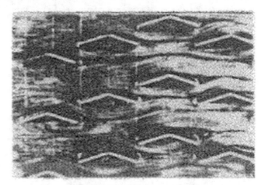

(b) Type II ($\theta = 17.4$ deg)

(c) Type II ($\theta = 24.6$ deg)

f factors for the OSF geometry. Figure 3.12 shows the visualization experiments of
OSF and OCLF arrays. Hitachi (1984), Fig. 3.13, shows the commercial use of the
OCLF geometry. Hatada and Senshu study indicated that the OCLF provides higher
performance than what is yielded by the standard OSF. Pauley and Hodgson (1994)

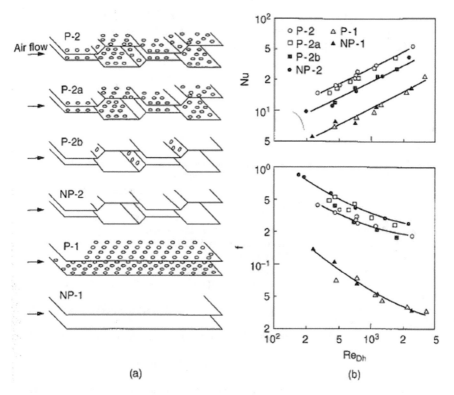

(a) (b)

Fig. 3.12 (a) Perforated plate geometries tested by Fujii et al. (1988), (b) test results on illustrated surfaces

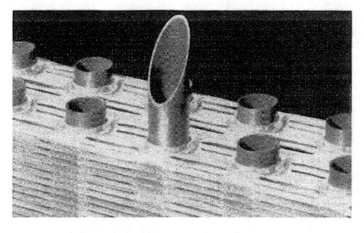

Fig. 3.13 The OCLF geometry in the Hitachi fin and tube heat exchanger (courtesy of Hitachi Cable)

visualized the flow. At low Reynolds number, the upstream half mass transfer coefficients are higher than those of the leeward half yield higher mass transfer coefficients, the reason being the unsteadiness in the shear layer impinging on the leeward surface.

References

Achaichia A, Cowell IA (1988) Heat transfer and pressure drop characteristics of flat tube and louvered plate fin surfaces. Exp Therm Fluid Sci 1:147–157

Aoki H, Shinagawa T, Suga K (1989) An experimental study of the local heat transfer characteristics in automotive louvered fins. Exp Therm Fluid Sci 2:293–300

Beecher DT, Fagan TJ (1987) Effects of fin pattern on the air-side heat transfer coefficient in plate finned-tube heat exchangers. ASHRAE Trans 93(2):1961–1984

Bellows KD (1996) Flow visualization of louvered-fin heat exchangers. In: Master's thesis. University of Illinois, Urbana, Champaign

Burgess NK, Ligrani PM (2004) Effects of dimple depth on Nusselt numbers and friction factors for internal cooling in a channel. In: ASME turbo expo 2004: power for land, sea, and air. ASME, pp 989–998

Chang YJ, Wang CC (1996) Air side performance of brazed aluminum heat exchangers. J Enhanc Heat Transf 3:15–28

Chang YJ, Wang CC (1997) A generalized heat transfer correlation for louver fin geometry. Int J Heat Mass Transfer 40:533–544

Chang YJ, Hsu KC, Lin YT, Wang CC (2000) A generalized friction correlation for louver fin geometry. Int J Heat Mass Transfer 43:2237–2243

Chyu MK, Yu Y, Ding H, Downs JP, Soechting FO (1997) Concavity enhanced heat transfer in an internal cooling passage. In: ASME 1997 international gas turbine and aeroengine congress and exhibition. ASME, pp. V003T09A080–V003T09A080

Critoph RE, Holland MK, Turner L (1996) Contact resistance in air-cooled plate fin-tube air-conditioning condensers. Int J Refrig 19:400–406

Critoph RE, Holland MK, Fisher M (1999) Comparison of steady state and transient methods for measurement of local heat transfer in plate fin-tube heat exchangers using liquid crystal thermography with radiant heating. Int J Heat Mass Transfer 42:1–12

Davenport CJ (1980) Heat transfer and fluid flow in the louvered-fin heat exchanger. PhD thesis, Lanchester Polytechnic, Lanchester, UK

Davenport CJ (1983a) Correlations for heat transfer and flow friction characteristics of louvered fin. In: Heat transfer Seattle, AIChE symposium, series no. 225(79):19–27

Davenport CJ (1983b) Heat transfer and flow friction characteristics of louvered heat exchanger surfaces heat exchangers: theory and practice. In: Taborek J, Hewitt GF, Afgan N (eds) . Hemisphere, Washington, DC, pp 387–412

DeJong NC, Jacobi AM (2003) Localized flow and heat transfer interactions in louvered-fin arrays. Int J Heat Mass Transfer 46(3):443–455

Dillen ER, Webb RL (1994) Rationally based heat transfer and friction correlations for the louver fin geometry. SAE Paper 940504, Warrendale, PA

Eckels PW, Rabas TJ (1987) Dehumidification: on the correlation of wet and dry transport processes in plate finned-tube heat exchangers. J Heat Transfer 109:575–582

Ekkad SV, Nasir H (2003) Dimple enhanced heat transfer in high aspect ratio channels. J Enhanc Heat Transf 10(4):395–406

Elyyan MA, Tafti DK (2009) Flow and heat transfer characteristics of dimpled multilouvered fins. J Enhanc Heat Transf 16(1):43–60

Fujii M, Seshimo Y, Yarnananaka G (1988) Heat transfer and pressure drop of the perforated surface heat exchanger with passage enlargement and contraction. Int J Heat Mass Transfer 31:135–142

Fujikake K, Aoki H, Mitui H (1983) An apparatus for measuring the heat transfer coefficients of finned heat exchangers by use of a transient method. In: Proc Japan 20th symposium on heat transfer, pp 466–468

Ha S, Kim C, Ahn S, Dreitser GA (1998) Condensate drainage characteristics of plate fin-and-tube heat exchanger. In: Heat exchangers for sustainable dev, Lisbon, Portugal, pp 423–430

Hatada T, Senshu T (1984) Experimental study on heat transfer characteristics of convex louver fins for air conditioning heat exchangers. ASME paper 84-HT-74, New York

Hitachi (1984) Chapter 5: Plate-and-fin extended surfaces. In: Hitachi high-performance heat transfer tubes cat no EA-500. Hitachi Cable Co., Tokyo, Japan, p 140

Huang CH, Yuan IC, Ay H (2003) A three-dimensional inverse problem in imaging the local heat transfer coefficients for plate finned-tube heat exchangers. Int J Heat Mass Transfer 46:3629–3638

Joshi HM, Webb RL (1987) Prediction of heat transfer and friction in the offset-strip fin array. Int J Heat Mass Transfer 30:69–84

Kajino M, Hiramatsu M (1987) Research and development of automotive heat exchangers. In: Heat transfer in high technology and power engineering conference, pp 420–432

Kayansayan N (1993) Heat transfer characterization of plate fin-tube heat exchangers. Heat Recov Syst CHP 13(1):67–68

Kays WM, Crawford ME (1980) Convective heat and mass transfer. McGraw-Hill, New York, p 151

Kim J-Y, Song T-H (2003) Effect of tube alignment on the heat/mass transfer from a plate fin and two-tube assembly: naphthalene sublimation results. Int J Heat Mass Transf 46:3051–3059

Kim NH, Yun JH, Webb RL (1997) Heat transfer and friction correlations for wavy plate fin-and-tube heat exchangers. J Heat Transfer 119(3):560–567

Kundu B, Das PK (1997) Optimum dimensions of plate fins for fin-tube heat exchangers. Int J Heat Fluid Flow 18:530–537

Kushida G, Yamashita H, Izumi R (1986) Fluid flow and heat transfer in a plate-fin and tube heat exchanger (analysis of heat transfer around a square cylinder situated between parallel plates). Bull JSME 29(258):4185–4191

Leu J-S, Wu Y-H, Jang J-Y (2004) Heat transfer and fluid flow analysis in plate-fin and tube heat exchangers with a pair of block shape vortex generators. Int J Heat Mass Transf 47:4327–4338

Ligrani PM, Harrison JL, Mahmmod GI, Hill ML (2001) Flow structure due to dimple depressions on a channel surface. Phys Fluids 13(11):3442–3451

Lyman AC, Stephan RA, Thole KA, Zhang LW, Memory SB (2002) Scaling of heat transfer coefficients along louvered fins. Exp Therm Fluid Sci 26(5):547–563

Mahmood GI, Hill ML, Nelson DL, Ligrani PM, Moon HK, Glezer B (2001) Local heat transfer and flow structure on and above a dimpled surface in a channel. J Turbomach 123(1):115–123

Maltson JD, Wilcock D, Davenport CJ (1989) Comparative performance of rippled fin plate fin and tube heat exchangers. J Heat Transfer 111:21–28

McQuiston FC (1978) Correlation of heat, mass, and momentum transport coefficients for plate-fin-tube heat transfer for surfaces with staggered tube. ASHRAE Trans 54(1):294–309

Michna GJ, Jacobi AM, Burton RL (2007) An experimental study of the friction factor and mass transfer performance of an offset-strip fin array at very high Reynolds numbers. J Heat Transfer 129(9):1134–1140

Moon HK, O'connell T, Glezer B (1999) Channel height effect on heat transfer and friction in a dimpled passage. In: ASME 1999 international gas turbine and aeroengine congress and exhibition. ASME, pp. V003T01A043–V003T01A043

Muzychka YS, Kenway G (2009) A model for thermal-hydraulic characteristics of offset strip fin arrays for large Prandtl number liquids. J Enhanc Heat Transf 16(1):73–92

Muzychka YS, Yovanovich MM (2001) Modeling the f and j characteristics of the offset strip fin array. J Enhanc Heat Transf 8(4):261–277

Nacerbey M, Russell S, Baudoin B (2003) PIV visualizations of the flow structure upstream of the tubes in a two-row plate fin-and-tube heat exchanger. In: Shah RK, Deakin AW, Honda H, Rudy TM (eds) Proc of the fourth int conf on compact heat exchangers and enhancement technology for the process industries. Begell House Inc., New York, pp 63–68

Pauley LL, Hodgson JE (1994) Flow visualization of convex louver fin arrays to determine maximum heat transfer conditions. Exp Therm Fluid Sci 9(1):53–60

Rich DG (1975) Effect of the number of tube rows on heat transfer performance of smooth plate fin-and-tube heat exchangers. ASHRAE Trans 81(1):307–319

Rugh JP, Pearson JT, Ramadhyani S (1992) A study of a very compact heat exchanger used for passenger compartment heating in automobiles in compact heat exchangers for power and process industries. ASME Symp Ser HTD 201:15–24

Saboya FEM, Sparrow EM (1974) Local and average heat transfer coefficients for one-row plate fin and tube heat exchanger configurations. J Heat Transfer 96:265–272

Saboya FEM, Sparrow EM (1976) Transfer characteristics of two-row plate fin and tube heat exchanger configurations. Int J Heat Mass Transf 19:41–49

Saboya FEM, Rosman EC, Carajilescov P (1984) Performance of one- and two-row tube and plate fin heat exchangers. J Heat Transfer 106:627–632

Sahnoun A, Webb RL (1992) Prediction of heat transfer and friction for the louver fin geometry. J Heat Transfer 114:893–900

Seshimo Y, Fujii M (1987) Heat transfer and friction performance of plate fin and tube heat exchangers at low Reynolds number (1st report: characteristic of single-row). Abstract Bull JSME 30:1688–1689

Seshimo Y, Fujii M (1991) An experimental study on the performance of plate fin and tube heat exchangers at low Reynolds numbers. In: Proc of the ASME-JSME thermal eng joint conference, vol 4, pp 449–454

Shah RK, Sekulic DP (2003) Fundamentals of heat exchanger design. John Wiley & Sons, New York

Shah RK, Webb RL (1982) Compact and enhanced heat exchangers. In: Taborek J, Hewitt GF, Afgan NH (eds) Heat exchangers: theory and practice. Hemisphere, Washington, DC, pp 425–468

Suga K, Aoki H (1995) Numerical study on heat transfer and pressure drop in multilouvered fins. J Enhanc Heat Transf l(3):231–238

Suga T, Aoki H (1991) Numerical study on heat transfer and pressure drop in multilouvered fins. In: Proc ASME/JSME joint thermal engineering conference, vol 4

Suga K, Aoki H, Shinagawa T (1990) Numerical analysis on two dimensional flow and heat transfer of louvered fins using overlaid grids. JSME Int J Ser II 33:122–127

Sunden B, Svantesson J (1990) Thermal hydraulic performance of new multilouvered fins. In: Proc 9th int heat trans conf, vol 5, pp 91–96

Suzuki K, Hirai E, Miyake T, Sato T (1985) Numerical and experimental studies on a two-dimensional model of an offset-strip-fin type compact heat exchanger used at low Reynolds number. Int J Heat Mass Transfer 28(4):823–836

Tafti DK (1999) Time-dependent calculation procedure for fully developed and developing flow and heat transfer in louvered fin geometries. Numer Heat Transfer Part A Appl 35(3):225–249

Tafti DK, Zhang X (2001) Geometry effects on flow transition in multilouvered fins–onset, propagation, and characteristic frequencies. Int J Heat Mass Transfer 44(22):4195–4210

Taler D (2004) Experimental determination of heat transfer and friction correlations for plate fin-and-tube heat exchangers. J Enhanc Heat Transf 11:183–204

Tanaka T, Itoh M, Kudoh M, Tomita A (1984) Improvement of compact heat exchangers with inclined louvered fins. Bull JSME 27(224):219–226

Tomoda T, Suzuki K (1988) A numerical study of heat transfer on compact heat exchanger (effect of fin shape). In: 25th national heat transfer symp. of Japan, pp 175–177

Torikoshi K, Xi GN, Nakazawa Y, Asano H (1994) Flow and heat transfer performance of a plate fin-and-tube heat exchanger (1st report: effect of fin pitch). In: Heat trans proc of the 10th int heat trans conf, vol 4, pp 411–416

Toyoshima S, Fukumoto H, Nakagawa Y, Sakamoto Y (1986) Numerical analysis on flow of plate-fin heat exchangers. In: Proceedings, 246th Lecture meeting of The Japan Society of Mechanical Engineers Kansai Branch, pp 864–1

Tutar M, Akkoca A (2004) Numerical analysis of fluid flow and heat transfer characteristics in three-dimensional plate fin-and-tube heat exchangers. Numer Heat Transfer Part A 46:301–321

Wang C-C, Lee C-J, Chang C-T, Lin S-P (1999) Heat transfer and friction correlation for compact louvered fin-and-tube heat exchangers. Int J Heat Mass Transf 42:1945–1956

Wang CC, Chang CT (1998) Heat and mass transfer for plate fin-and-tube heat exchangers with and without hydrophilic coating. Int J Heat Mass Transfer 41:3109–3120

Wang CC, Chang YJ, Hsieh YC, Lin YT (1996) Sensible heat and friction characteristics of plate fin-and-tube heat exchangers having plane fins. Int J Refrig 19(4):223–230

Webb RL (1988) PSU unpublished data for five radiators

Webb RL, Jung SH (1992) Air-side performance of enhanced brazed aluminum heat exchangers. ASHRAE Trans 98(2):391–401

Webb RL, Trauger P (1991) How structure in the louvered fin heat exchanger geometry. Exp Therm Fluid Sci 4(2):205–217

Webb RL, Chang YJ, Wang CC (1995) Heat transfer and friction correlations for the louver fin geometry. In: IMechE symp C496/081/95

Webb RL, Kim NY (2005) Principles of enhanced heat transfer. Taylor and Francis, New York

Yamashita H, Kushida G, Izumi R (1987) Fluid flow and heat transfer in a plate-fin and tube heat exchanger (analysis of unsteady flow and heat transfer around a square cylinder situated between parallel plates. In: Proc of the 1987 ASME-JSME thermal eng joint conf, vol 4, pp 173–180

Zhang X, Tafti DK (2001) Classification and effects of thermal wakes on heat transfer in multilouvered fins. Int J Heat Mass Transfer 44(13):2461–2473

Zhang X, Tafti DK (2003) Flow efficiency in multi-louvered fins. Int J Heat Mass Transfer 46:1737–1750

Chapter 4
Vortex Generators

Streamwise vortices are shed from geometric shapes attached to the wall when the surfaces attack the flow at an angle. Vortex generators of different types have been investigated by Fiebig et al. (1993). Fundamental aspects of vortex generators and vortex generators applied to the plate-and-fin geometry are to be understood for heat transfer enhancement.

When the protrusion height is comparable to the local boundary layer thickness, protrusions are called vortex generators. The vortices bend around the protrusions; these are carried downstream in a longitudinal vortex pattern.

Longitudinal vortices were found by Eibeck and Eaton (1987). The protrusions having heights higher than adequate will cause a significantly higher pressure drop increase than the heat transfer enhancement. Arrangements of vortex generators are shown in Fig. 1.3. Mehta et al. (1983) have investigated the interaction of adjacent vortices divided into two types: common flow down and common flow up. When the direction of the secondary flow between two counter-rotating vortices is towards the wall or away from the wall, the vortices are called common flow down or common flow up.

Torii et al. (1994) investigated longitudinal vortices behind a vortex generator placed in a laminar boundary layer on a flat plate, Fig. 1.4. The vortex is formed by the flow separation on the leading edge of the wing and the corner vortex is due to deformation of near-wall vortex lines at the pressure side of the wing. Sometimes, also a vortex may be induced opposite to the main and corner vortices. Figure 1.5, Pauley and Eaton (1988) shows the contour plot of the streamwise velocity. Boundary layer thinning in respective cases of upward or downward is responsible for fluid mixing, momentum, and thermal energy diffusion and enhancement.

Fiebig (1995) has studied extensively the effects of size, shape, angle of attack, aspect ratio, etc. and Reynolds number on the heat transfer enhancement by vortex generators. Vortex generators are particularly useful for laminar flow heat transfer enhancement with limited increase in pressure drop. Fiebig (1995) worked with the vortex generators having heights comparable to the channel height. Liquid crystal

© The Author(s), under exclusive license to Springer Nature Switzerland AG 2020   79
S. K. Saha et al., *Heat Transfer Enhancement in Plate and Fin Extended Surfaces*,
SpringerBriefs in Applied Sciences and Technology,
https://doi.org/10.1007/978-3-030-20736-6_4

thermography and drag force measurements were used to find heat transfer coefficients and pressure drops, respectively.

Fiebig (1995) characterized the compact heat exchanger by using wing-type vortex generator (WVG). Different WVGs were evaluated experimentally and numerically to enhance the heat transfer and reduce the pressure loss. He presented the comparison of WVG fins with offset-strip fins and louvered fins. Figure 4.1 shows the comparison between offset-strip fin and louvered fin configuration reported in Kays and London (1984) with the delta wings (DW) of Brockmeier et al. (1989) and the ISB configuration with $\beta = 15°$ and $h/H = 0.2$ for $500 \le Re \le 2000$. Vortex generators in heat exchanger influenced the heat transfer (Milliat 1961; Edwards and Sherill 1974; Russel et al. 1982; Fiebig et al. 1989; Fiebig and Guntermann 1989; Amon 1989; Dong 1989; Zhang 1989; Tiggelbeck 1990; Riemann 1992; Valencia 1993; Fiebig and Guntermann 1993a, b). The louvered fin and offset-strip fin had basically same flow structure and working principle but difference was that the flow meanders through the louvers. Figure 4.2 shows the schematic diagram of offset-strip fins and louvered fins.

Types of vortex generators:

1. *Transverse vortex generator (TVG)*: It is the channel with ribs and corrugations (Amon and Mikic 1989, 1990; Greiner et al. 1989; Ellouze et al. 1993; Grosse-Gorgemann et al. 1993b, c; Herman et al. 1991). It generates swirl with axis of rotation perpendicular to the main flow direction. Figure 4.3 shows periodic

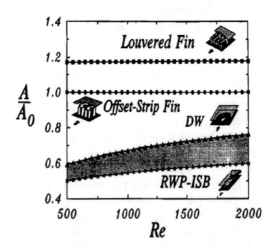

Fig. 4.1 Heat transfer surface requirement for different fins, Guntermann (1992). Reference surface: Offset-strip fin, mass flow ratio = 1, heat duty ratio = 1, pumping power ratio = 1, hydraulic diameter ratio = 1, temperature difference ratio = 1; data: louvered fin:3/16–11.1, Kays and London (1984), offset-strip fin: 3/32–12.22, Kays and London (1984); delta wing fin: $L_P = 10H$, $B = 4H$, $\beta = 30°$, $I = 2H$; ISB fin: $L_P = 3.86$, $B_P = 1.03H$, $\beta = 15°$, $I = 2H$, $h = 0.2H$, $Pr = 0.7$, $T_W = 2T_O$ at the entrance of periodic element (Fiebig 1995)

Offset-Strip Fin   **Louvered Fins**

Fig. 4.2 Schematic diagram of high-performance fins (Fiebig 1995)

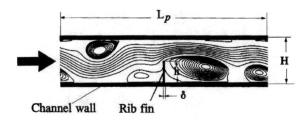

Channel wall   Rib fin

Fig. 4.3 Instantaneous streamlines of transverse vortex street, periodic in time and space (Karman vortex street), generated by TVGs in the form of periodic fin ribs in a channel (Grosse-Gorgemann et al. 1993)

transverse vortex street of a periodically ribbed channel. Unsteady flow is required for transverse vortices and that creates reversed flow regions.

2. *Longitudinal vortex generator (LVG)*: Longitudinal vortices are more efficient than transverse vortices for heat transfer enhancement. Strong swirling mechanism is required to enhance the heat transfer for longitudinal vortices.

3. *Wing-type vortex generator (WVG)*: WVG found two forms, delta form and rectangular form. Wing and winglet VGs, mounted and punched WVGs, single rows, double rows, and periodic arrays of WVGs had been investigated by varying angle of attack (Fiebig et al. 1986; Riemann 1992), aspect ratio (Fiebig et al. 1986; Tiggelbeck 1990; Chen 1993; Zhang 1989; Brockmeier 1987), winglet inclination to primary surface (Tiggelbeck 1990), in-line and staggered arrangement (Tiggelbeck et al. 1993), parallel winglet configurations (Riemann 1992; Tiggelbeck 1990), and flow reversal (Kallweit 1986).

Vortex generators are effective over an area several hundred times the vortex generator area. The effectiveness of the vortex generators depends on the ratio of vortex generator area to the primary surface area. Delta-type vortex generators are more useful than rectangular forms. Winglets are better than wings. Heat transfer increases with the angle of attack up to maximum angle which depends on the form of vortex generator, A_{VG}/A and Re. Pressure drop increases monotonically with the angle of attack, A_{VG}/A and Re. Pressure loss increase is mainly due to form drag of the vortex generators. Transition to turbulence occurs at smaller Reynolds number

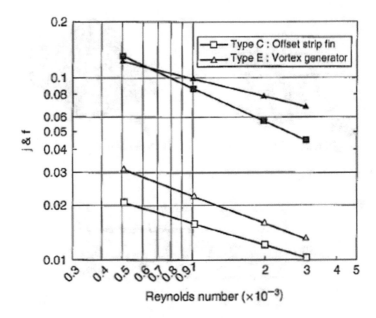

Fig. 4.4 The j and f factor of a channel mounted with vortex generator (from numerical study) compared with OSFs (Kays and London 1984, reported by Brockmeier et al. 1993)

than in plain channel flow. Vortex breakdown is peculiar to external flow only. This does not occur in constricted internal flow.

The effect of longitudinal spacing between rows in multirow configuration of vortex generators was investigated by Tiggelbeck et al. (1994) for a two-row configuration. Brockmeier et al. (1993), Fig. 4.4, performed a numerical simulation on the delta winglet vortex generators applied to a plate-fin geometry.

Li et al. (2017) investigated the thermo-hydrodynamic characteristics of a pin-fin heat sink with delta winglet vortex generators in a crossflow by both experimentally and numerically. They studied the effect of Reynolds number of the flow and the geometrical parameters of the vortex generator on the performance of the heat sink. They observed that thermal resistance decreases as the Reynolds number increases. The angle of attack of 30° was better arrangement for vortex generator by considering thermal resistance and pressure difference simultaneously. They also observed that heat transfer can be increased by installing the vortex generator at the middle of the heat sink and increasing the height of the vortex generator.

Kotcioglu and Caliskan (2008) experimentally studied the heat transfer and fluid flow characteristics in a plate-fin heat exchanger which was configured with wing-type convergent-divergent longitudinal vortex generators (CDLVGs). The objective of this study was to compute heat transfer and pressure drop characteristic of CDLVGs as well as to evaluate the effectiveness of plate-fin heat exchanger operated under different conditions. They considered both cold flow and hot flows and used rectangular fins.

They conducted experiments for the following conditions: (a) fully opened flow clack, (b) half-opened hot flow clack and fully opened cold flow clack, and (c) half-opened cold flow clack and fully opened hot flow clack. They experimented precisely and for every run calculated pressure drop across orifice plates, pressure drop at inlet and outlet of the test section, surrounding temperature, and temperature at inlet and outlet of heat exchanger. The winglet approach angle was $\beta_h = 30°$ for hot fluid and $\beta_c = 60°$ for cold fluid in the direction of flow. Uncertainties were as per the method proposed by Kline and McClintok (1953), Kotcioğlu et al. (1998), and Oğulata et al. (2000). The inter-fin distance is helpful in estimating the thermal resistance, flow velocity, angle of attack, and dimensionless pressure drop as Sahin et al. (2005) and Yakut et al. (2006) considered. Kotcioglu and Caliskan (2008) established dimensionless correlations for Nusselt number and pressure drop coefficient as

$$Nu = aRe^b \left(\frac{L_a}{L_b}\right)^c \left[\frac{c}{L_c}\right]^d (\tan \beta)^g \qquad (4.1)$$

and friction factor correlations as

$$f = C_o Re^{-m} \qquad (4.2)$$

They concluded simultaneous increase in Nusselt number and friction factor as expected. The CDLVG element enhances heat transfer up to 120%. It also increases both pressure drop and friction factor. The plate-fin array system was evaluated by ε-NTU technique. NTU variation was found to be between 3.32 and 3.85 and they found 68–80% increase in effectiveness with the CDLVG. Also, they observed a second case where half-opened flow clack for hot flow and fully opened flow clack for cold flow were best performers in the heat transfer enhancement. Final conclusion was that secondary flows are important in heat transfer; thus, in the form of nozzle or venturi, the vortex generator can be designed.

Brockmeier et al. (1989), Esformes (1989), Fiebig et al. (1995), Lee (1979), Lee and Kwon (1992), Mullisen and Loehrke (1986), Pang et al. (1990), Patankar and Prakash (1981), Pescod (1974), Tauscher and Mayinger (1997), and Vasudevan et al. (2000) studied the heat transfer and pressure drop characteristics of vortex generators.

Li et al. (2013) studied the performance of a pair of vortex generators to enhance the heat transfer rate from a flat-fin heat sink. The delta winglet vortex generators were used. The effect of various parameters of the vortex generators such as angle of attack, height, distance between their trailing edges, and distance between the trailing edge of the delta winglet vortex generator and the heat sink on the heat transfer and pressure drop characteristics has been presented. The results of the above-said effects have been obtained by using infrared thermography. The analysis has been carried out under crossflow condition.

They observed that the thermal resistance to heat transfer from the heat sink decreased due to the presence of vortex generators. But the rate at which the thermal resistance decreases is depleted with increase in the distance between the trailing edges of the vortex generators. The thermal resistance to flow over heat sink with vortex generators is greater than that over heat sink without vortex generators because they act as an obstacle to the flow and result in flow bypass. However, when the distance between the trailing edges is greater than or equal to the length of the heat sink then the effect of fluid mixing is stronger and helps in reducing the thermal resistance. They reported that the heat transfer rate was most efficient from energy consumption point of view, when the vortex generators are in alignment with the heat sink. Also, 30° angle of attack and H height resulted in better heat transfer efficiency.

Wang et al. (2002), Leu et al. (2004), Chen and Shu (2004), Pesteei et al. (2005), Ferrouillat et al. (2006), Chomdee and Kiatsiriroat (2006, 2007), Joardar and Jacobi (2007), Tian et al. (2009), Yang et al. (2010a, b), Henze et al. (2011a, b), Ahmed et al. (2012a, b), Huisseune et al. (2013a, b), Sinha et al. (2013), Dake and Majdalani (2009), Min et al. (2010), Aris et al. (2011), and Althaher et al. (2012) have studied the performance of different vortex generators.

References

Ahmed HE, Mohammed HA, Yusoff MZ (2012a) An overview on heat transfer augmentation using vortex generators and nanofluids: approaches and applications. Renew Sust Energ Rev 16:5951–5993

Ahmed HE, Mohammed HA, Yusoff MZ (2012b) Heat transfer enhancement of laminar nanofluids flow in a triangular duct using vortex generators. Superlattice Microst 52:398–415

Althaher MA, Abdul-Rassol AA, Ahmed HE, Mohammed HA (2012) Turbulent heat transfer enhancement in a triangular duct using delta-winglet vortex generators. Heat Transfer Asian Res 41:43–62

Amon CH (1989) Numerical investigation of starting flow and supercritical heat transfer enhancement in grooved channels: understanding and exploitation. In: Proceedings of the 10th Brazil cong mech eng, Rio de Janeiro, pp 197–200

Amon CH, Mikic BB (1989) Spectral element simulation of forced convective heat transfer. Application to slotted channel flow. In: National heat transfer conference, HTD, vol 110, pp 175–183

Amon CH, Mikic BB (1990) Numerical prediction of convective heat transfer in self-sustained oscillatory flows. J Thermophys Heat Transfer 4(2):239–246

Aris MS, McGlen R, Owen I, Sutcliffe CJ (2011) An experimental investigation into the deployment of 3-D, finned wing and shape memory alloy vortex generators in a forced air convection heat pipe fin stack. Appl Therm Eng 31:2230–2240

Brockmeier U (1987) Numerisches Verfahren zur Berechnung dreidimensionaler Stromungs- und Temperaturfelder in Kanlilen mit Llingswirbelerzeugern und Untersuchung von Warmeiibergang und Stromungsverlust. Dissertation, Ruhr-Universitiit Bochum

Brockmeier U, Fiebig M, Güntermann T, Mitra NK (1989) Heat transfer enhancement in fin-plate heat exchangers by wing type vortex generators. Chem Eng Technol 12(1):288–294

Brockmeier U, Guntermann T, Fiebig M (1993) Performance evaluation of a vortex generator heat transfer surface and comparison with different high performance surface. Int J Heat Mass Transfer 36:2575–2587

Chen Y (1993) Numerische Untersuchungen von Lamellen-RohrWarmeiibertragerelementen unter Beriicksichtigung der Warmeleitung in den Lamellen. Diplomarbeit Nr. 93/12, Ruhr-Universitiit Bochum

Chen TY, Shu HT (2004) Flow structures and heat transfer characteristics in fan flows with and without delta-wing vortex generators. Exp Thermal Fluid Sci 28:273–282

Chomdee S, Kiatsiriroat T (2006) Enhancement of air cooling in staggered array of electronic modules by integrating delta winglet vortex generators. Int Commun Heat Mass Transfer 33:618–626

Dake T, Majdalani J (2009) Improving flow circulation in heat sinks using quadrupole vortices. In: Proceedings of the ASME 2009 InterPACK conference. American Society of Mechanical Engineers, San Francisco, CA

Dong Y (1989) Experimentelle Untersuchung der Wechselwirkungen von Liingswirbelerzeugern und Kreiszylindern in Kanalstromungen in Bezug aufWarmeiibergang und Stromungsverlust. Dissertation, Ruhr-Universitiit Bochum

Edwards FJ, Sherill N (1974) The improvement of forced surface heat transfer using surface protrusions in the form of cubes and vortex generators. In: Proceedings of the 5th international heat transfer conference, vol 2. Tokyo, pp 244–248

Eibeck PA, Eaton JK (1987) Heat transfer effects of a longitudinal vortex embedded in a turbulent shear flow. J Heat Transfer 109:16–24

Ellouze A, Blancher S, Crelf R (1993) Flow structure and heat transfer in a wavy wall channel at steady and unsteady flow regime. In: Proc Eurotherm 31 "Vortices and Heat Transfer", Bochum, Germany, pp 30–35

Esformes JL (1989) Ramp wing enhanced plate fin. U.S. patent 4,817, p 709

Fiebig M (1995) Vortex generators for compact heat exchangers. J Enhanc Heat Transf 2:1–2

Fiebig M, Brockmeier U, Mitra NK, Gü Termann T (1989) Structure of velocity and temperature fields in laminar channel flows with longitudinal vortex generators. Numer Heat Transfer Appl 15(3):281–302

Fiebig M, Giintermann T (1989) Heat transfer enhancement by longitudinal vortex generators. In: Proceedings of the 10th Brazil cong mech eng, Rio de Janeiro, pp 445–448

Fiebig M, Valencia A, Mitra NK (1993) Wing-type vortex generators for fin-and-tube heat exchangers. Exp Therm Fluid Sci 7(4):287–295

Fiebig M, Kallweit P, Mitra NK (1986) Wing type vortex generators for heat transfer enhancement. IHTC, vol 6, pp 2909–2913

Fiebig M, Guntermann T (1993a) A class of high performance compact fin-plate heat exchanger elements. In: Lee JS, Chung SH, Kim KH (eds) The 6th Int symp on transport phenomena in thermal engineering, vol III. Korean Society of Mechanical Engineering, Seoul, pp 49–54

Fiebig M, Guntermann T (1993b) Heat transfer surfaces with longitudinal vortex generators for compact plate heat exchangers. In: Proc 1st international thermal energy congress ITEC93, vol 1. Marakesch

Fiebig M, Guntermann T, Mitra NK (1995) Numerical analysis of heat transfer and flow loss in a parallel plate heat exchanger element with longitudinal vortex generators as fins. J Heat Transfer 117(4):1064–1068

Ferrouillat S, Tochon P, Garnier C, Peerhossaini H (2006) Intensification of heat-transfer and mixing in multifunctional heat exchangers by artificially generated streamwise vorticity. Appl Therm Eng 26:1820–1829

Guntermann T (1992) Dreidimensionale stationare und selbsterregt-schwingende Stromungs- und Temperaturfelder in Hochleistungswiirmeiibertragern mit Wirbelerzeugern. Dissertation, RuhrUniversitat Bochum

Greiner M, Chen RF, Witz RA (1989) Heat transfer augmentation through wall shape induced flow destabilization. In: National heat tranefer conference, HTD, vol 107

Grosse-Gorgemann A, Weber D, Fiebig M (1993b) Numerical and experimental investigation of self-sustained oscillations in channels with periodic structures. In: Proc Eurotherm 31 "Vortices and Heat Transfer", Bochum, Germany, pp 42–50

Grosse-Gorgemann A, Weber D, Fiebig M (1993c) Self-sustained oscillations: heat transfer and flow losses in Laminar channel flow with rectangular vortex generators. In: Proc Eurotherm 31 "Vortices and Heat Transfer", Bochum, Germany, pp 107–111

Henze M, von Wolfersdorf J, Weigand B, Dietz CF, Neumann SO (2011) Flow and heat transfer characteristics behind vortex generators – a benchmark dataset. Int J Heat Fluid Flow 32:318–328

Henze M, von Wolfersdorf J (2011) Influence of approach flow conditions on heat transfer behind vortex generators. Int J Heat Mass Transf 54:279–287

Herman CV, Mayinger F, Sekulic DP (1991) Experimental verification of oscillatory phenomena in heat transfer in a Communicating Channel geometry. Proc 2nd world conf on exp heat transf, Fluid mech and thermodynamics, June 23–28, Dubrovnik, Yugoslavia

Huisseune H, T'Joen C, De Jaeger P, Ameel B, De Schampheleire S, De Paepe M (2013a) Performance enhancement of a louvered fin heat exchanger by using delta winglet vortex generators. Int J Heat Mass Transf 56:475–487

Huisseune H, T'Joen C, De Jaeger P, Ameel B, De Schampheleire S, De Paepe M (2013b) Influence of the louver and delta winglet geometry on the thermal hydraulic performance of a compound heat exchanger. Int J Heat Mass Transf 57:58–72

Joardar A, Jacobi AM (2007) A numerical study of flow and heat transfer enhancement using an array of delta-winglet vortex generators in a fin-and-tube heat exchanger. J Heat Transf 129:1156–1167

Kallweit P (1986) Liingswirbelerzeuger fiir den Einsatz in Lamellenwiirmetauschern. Dissertation, Ruhr-Universitiit Bochum

Kays WM, London AL (1984) Compact heat exchangers. 3rd Edition, McGraw-Hill, New York

Kline SJ, McClintok F (1953) Describing uncertainty in single sample experiments. Mech Eng 75:3–8

Kotcioglu I, Caliskan S (2008) Experimental investigation of a cross-flow heat exchanger with wing-type vortex generators. J Enhanc Heat Transf 15(2):113–127

Kotcioğlu İ, Ayhan T, Olgun H, Ayhan B (1998) Heat transfer and flow structure in a rectangular channel with wing-type vortex generator. Turk J Eng Environ Sci 22(3):185–196

Lee GH (1979) Effect of vortex generators on the heat transfer from rectangular plate fins. The Lumus Company Limited, Heat Transfer Division, England, Report No. HR-159

Lee KB, Kwon YK (1992) Flow and thermal field with relevance to heat transfer enhancement of interrupted-plate heat exchangers. Exp Heat Transfer 5:83–100

Leu JS, Wu YH, Jang JY (2004) Heat transfer and fluid flow analysis in plate-fin and tube heat exchangers with a pair of block shape vortex generators. Int J Heat Mass Transf 47:4327–4338

Li HY, Chen CL, Chao SM, Liang GF (2013) Enhancing heat transfer in a plate-fin heat sink using delta winglet vortex generators. Int J Heat Mass Transf 67:666–677

Li HY, Liao WR, Li TY, Chang YZ (2017) Application of vortex generators to heat transfer enhancement of a pin-fin heat sink. Int J Heat Mass Transf 112:940–949

Mehta RD, Shabaka IM, Shibi A, Bradshaw P (1983) Longitudinal vortices imbedded in turbulent boundary layers. AIAA Paper, Albuquerque, NM

Milliat JP (1961) Experimental study of finned cans of the 'herring-bone' type. In: Int. j. Brit. nuclear energy conf., vol 6, Electricite de France, Chatou

Min C, Qi C, Kong X, Dong J (2010) Experimental study of rectangular channel with modified rectangular longitudinal vortex generators. Int J Heat Mass Transf 53:3023–3029

Mullisen RS, Loehrke RI (1986) A study of the flow mechanisms responsible for heat transfer enhancement in interrupted-plate heat exchangers. J Heat Transfer 108:377–385

Oğulata RT, Doba F, Yilmaz T (2000) Irreversibility analysis of cross flow heat exchangers. Energy Convers Manag 41(15):1585–1599

Pang K, Tao WQ, Zhang HH (1990) Numerical analysis of fully developed fluid flow and heat transfer for arrays of interrupted plates positioned convergently-divergently along the flow direction. Numer Heat Transfer Part A 18:309–324

Patankar SV, Prakash C (1981) An analysis of the effect of plate thickness on laminar flow and heat transfer in interrupted plate passages. Int J Heat Mass Transfer 24:1801–1810

Pauley WR, Eaton JK (1988) Experimental study of the development of longitudinal vortex pairs embedded in a turbulent boundary layer. AIAA J 26:816–823

Pescod D (1974) The effects of turbulence promoters on the performance of plate heat exchangers. In: Heat exchangers: design and theory sourcebook. Scripta Book Company, Washington, pp 601–616

Pesteei SM, Subbarao PM, Agarwal RS (2005) Experimental study of the effect of winglet location on heat transfer enhancement and pressure drop in fin-tube heat exchangers. Appl Therm Eng 25 (11–12):1684–1696

Riemann K-A (1992) Wiirmeiibergang und Druckabfall in Kaniilen mit periodischen Wirbelerzeugern bei thermischem Anlauf. Dissertation, Ruhr-Universitiit Bochum

Russel CMB, Jones TV, Lee GH (1982) Heat transfer enhancement using vortex generators. In: Proceedings of the 7th international heat transfer conference, vol 3, pp 283–288

Sahin B, Yakut K, Kotcioglu I, Celik C (2005) Optimum design parameters of a heat exchanger. Appl Energy 82(1):90–106

Sinha A, Raman KA, Chattopadhyay H, Biswas G (2013) Effects of different orientations of winglet arrays on the performance of plate-fin heat exchangers. Int J Heat Mass Transf 57:202–214

Tauscher R, Mayinger F (1997) Enhancement of heat transfer in a plate heat exchanger by turbulence promoters. In: Shah RK, Bell KJ, Mochizuki S, Wadekar VW (eds) Proc of the int conf on compact heat exchangers for the process industries. Begell House Inc., New York, pp 253–360

Tian LT, He YL, Lei YG, Tao WQ (2009) Numerical study of fluid flow and heat transfer in a flat-plate channel with longitudinal vortex generators by applying field synergy principle analysis. Int Commun Heat Mass Transfer 36:111–120

Tiggelbeck S (1990) Experirnentelle Untersuchungen an Kanalstromungen mit Einzel- und Doppel-Wirbelerzeuger-Reihen fiir den Einsatz in kompakten Wiirmetauschem. Dissertation, RuhrUniversitiit Bochum

Tiggelbeck T, Mitra NK, Fiebig M (1993) Experimental investigations of heat transfer and flow losses in a channel with double rows of longitudinal vortex generators. Int J Heat Mass Transf 36(9):2327–2337

Tiggelbeck S, Mitra NK, Fiebig M (1994) Comparison of wing-type vortex generators for heat transfer enhancement in channel flows. J Heat Transfer 116:880–885

Torii K, Nishina K, Nakayama K (1994) Mechanism of heat transfer augmentation by longitudinal vortices in a flat plate boundary layer. In: Heat transfer proc 10th int heat trans conf, vol 5, pp 123–128

Valencia A (1993) Wiirmeiibergang und Druckverlust in LamellenRohr-Wiirmeiibertragern mil Liingswirbelerzeugern. Dissertation, Ruhr-Universitiit Bochum

Vasudevan R, Eswaran V, Biswas G (2000) Winglet-type vortex generators for plate-fin heat exchangers using triangular fins. Numer Heat Trans Part A 38(5):533

Wang CC, Lo J, Lin YT, Wei CS (2002) Flow visualization of annular and delta winlet vortex generators in fin-and-tube heat exchanger application. Int J Heat Mass Transf 45(18):3803–3815

Yakut K, Alemdaroglu N, Kotcioglu I, Celik C (2006) Experimental investigation of thermal resistance of a heat sink with hexagonal fins. Appl Therm Eng 26(17–18):2262–2271

Yang KS, Li SL, Chen IY, Chien KH, Hu R, Wang CC (2010a) An experimental investigation of air cooling thermal module using various enhancements at low Reynolds number region. Int J Heat Mass Transf 53:5675–5681

Yang KS, Jhong JH, Lin YT, Chien KH, Wang CC (2010b) On the heat transfer characteristics of heat sinks: with and without vortex generators. IEEE Trans Compon Packag Technol 33:391–397

Zhang Z (1989) Einflu8 von Deltafugel-Wirbelerzeugem auf Wiirmeiibergang und Druckverlust in Spaltstromungen. Dissertation, Ruhr-Universitiit Bochum

Chapter 5
Wavy Fin, 3D Corrugated Fin, Perforated Fin, Pin Fin, Wire Mesh, Metal Foam Fin, Packings, Numerical Simulation

Wave pitch (P_w), corrugation angle (θ), and channel spacing (s) are the main parameters of the wavy or corrugated fin geometry having constant corrugation angles and sharp wave tips and the parameters influence the hydrothermal performance. Moreover, sharp corners instead of smooth corners influence the performance.

Mithun Krishna et al. (2018) computed the performance of wavy microchannels in order to achieve increased heat transfer rates from electronic chips. The surfaces with waviness have been shown in Fig. 5.1. The heat transfer enhancement using wavy microchannels is mainly due to the breaking of the flow and destabilizing of it. Burns and Parkes (1967) were known to be the first to study wavy channel performance. Figure 5.2 presents the change in average Nusselt number and average friction factor with respect to Reynolds number for different aspect ratios.

They observed that both Nu_{avg} and f_{avg} increased with decrease in aspect ratio. Also, the channel effectiveness corresponding to different aspect ratios has been shown in Fig. 5.3. They have also proposed a hybrid microchannel. The design of the hybrid channel and its performance has been shown in Fig. 5.4. They reported that the increase in heat transfer rate and friction factor for wavy microchannel relative to those corresponding to straight microchannel are 130% and 35%, respectively. On the other hand, the performance of hybrid microchannels was observed to be inferior to that of wavy microchannels.

Goldstein and Sparrow (1977), Saniei and Dini (1993), Wang and Vanka (1995), Rush et al. (1999), Mahmud et al. (2002), Qu and Mudawar (2002), Islamoglu and Parmaksizoglu (2003), Comini et al. (2003), Metwally and Manglik (2004), Morini (2004), Lee et al. (2005), Rosaguti et al. (2006, 2007) Oviedo-Tolentino et al. (2008), Sui et al. (2010, 2011), Gong et al. (2011), Mohammed et al. (2011), Rostami et al. (2015), Abed et al. (2015), and Kirsch and Thole (2017) presented the advantages of wavy channels for heat transfer enhancement.

Goldstein and Sparrow (1977) studied the mass transfer by naphthalene sublimation technique and drew the analogy with the thermal problem. They used

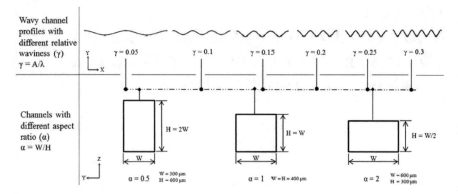

Fig. 5.1 Surface configurations with relative waviness (Mithun Krishna et al. 2018)

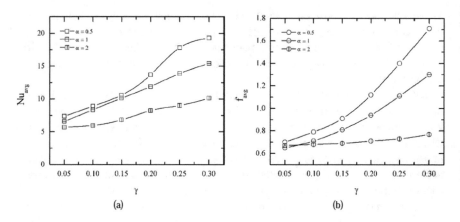

Fig. 5.2 Variation of (**a**) average Nusselt number and (**b**) average friction factor with Reynolds number (Mithun Krishna et al. 2018)

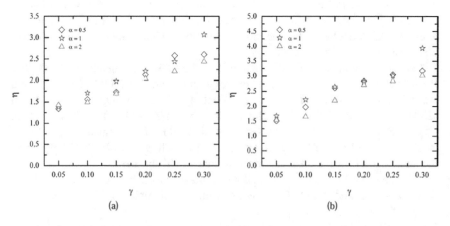

Fig. 5.3 Channel effectiveness corresponding to different aspect ratios (Mithun Krishna et al. 2018)

Fig. 5.4 Design of hybrid channel and its performance (Mithun Krishna et al. 2018)

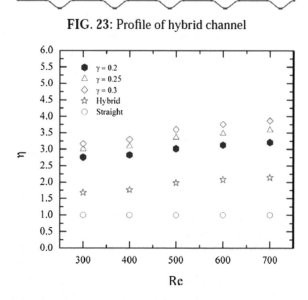

FIG. 23: Profile of hybrid channel

herringbone wave configuration. Formation of Gortler vortices passing over the concave wave surfaces causes enhancement (Gschwind et al. 1995). These vortices are counter-rotating and they have a cork-screw-like flow pattern. For a true wavy channel, flow separation and reattachment occur locally on the concave surfaces and these are revealed by flow visualization. The redevelopment of the boundary layer downstream of the reattachment point contributes to the heat transfer enhancement.

Yu et al. (2017) carried out an experiment on flow between two wavy plate fins and obtained analytic solutions of the Fanning friction factor and the Nusselt number for low Reynolds number. The geometry of the wavy plate fins was defined by fin spacing ($2H$), amplitude of waviness (a), and period length (L). Two-dimensional flow was taken between both wavy plate fins. The governing equations were solved by coordinate transformation in conjunction with perturbation method. The results obtained from analytical solution were closed agreement with numerical solutions.

Wang et al. (1997), Tao et al. (2007a, b), Xie et al. (2013), Jang and Chen (1997), Rush et al. (1999), Zhang et al. (2004), and Zhou et al. (2016) studied the effect of waviness and spacing of wavy plate fin on the heat transfer performance and pressure drop characteristics. They observed that friction factor monotonically increases with increase in dimensionless waviness (a/L). Nusselt number first increased up to peak value and then decreased as H/L increased. They presented a correlation of $(H/L)_{max}$ for which Nusselt number attained maximum value.

Tao et al. (2007a, b) numerically investigated the air-side performance of wavy fin surface. The objective of the study was to investigate the local Nusselt number variation and its distribution on the corresponding plain plate and wavy fin surface. They used numerical methods for objective function and for determining the design of efficient wavy fin surface. They compared their results with Xin et al. (1994) and

Wang et al. (2002) at various Reynolds numbers ranging from 500 to 4000 and observed that the mean deviation of the corresponding Nusselt number was 3.3%. The local Nusselt number distribution was calculated by

$$Nu_{i,j} = \frac{\dot{q}_{w,i,j} D}{k\left(T_{w,i,j} - T_{b,i,j}\right)} \tag{5.1}$$

The simulation results presented that the wavy fins have better heat transfer performance compared to that of plain plate fin at identical Reynolds number. The performance of wavy fins is more dominant as Reynolds number increases. The three simulated fin-and-tube heat exchangers have been represented in Fig. 5.5 and corresponding Nusselt number versus Reynolds number characteristic curves are plotted in Fig. 5.6.

It confirms that the performance of fin-A type was better. However, according to Shah and London (1978) proposed index *Nu/f* for fin surface evaluation has been plotted in Fig. 5.7 which depicted that fin-B type performance was better. Tao et al. (2007a, b) compared the three types of fins at identical pumping power and pressure drop and concluded that fin-B type performance was best at identical pressure drop whereas fin A and fin B type have the comparable results and both were significantly

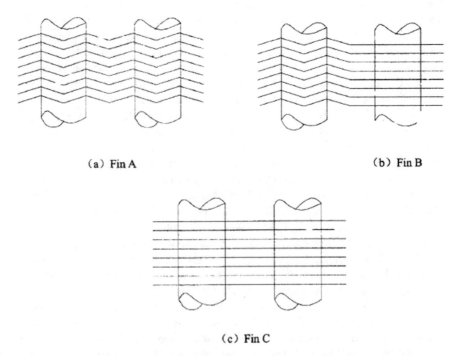

(a) Fin A   (b) Fin B

(c) Fin C

Fig. 5.5 Schematic diagram of the simulated three types of a fin-and-tube heat exchanger (Tao et al. 2007a, b)

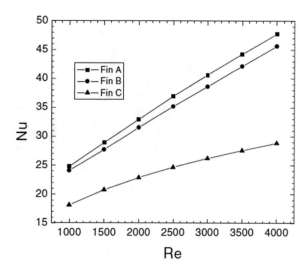

Fig. 5.6 Computational results of the Nusselt number against the Re number (Tao et al. 2007a, b)

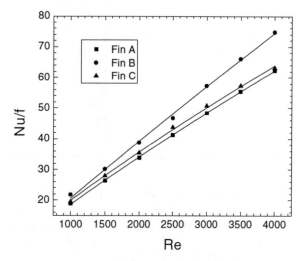

Fig. 5.7 Comparison of *Nu/f* against the Re number (Tao et al. 2007a, b)

better than type C fin. Tao et al. (2007a, b) concluded that heat transfer coefficient was significantly higher in upstream region than that of downstream region. Also, he observed that wave angle directs pressure drop.

Khoshvaght Aliabadi et al. (2014) presented correlation for heat transfer and pressure drop characteristics in a wavy plate-fin heat exchanger. The correlations have been developed using air, water, and ethylene glycol as working fluids. The wavy plate-fin and its geometrical details (fin height (F_h), fin pitch (F_p), wave length (L), fin amplitude (A), fin thickness (t), and fin length (F_D)) have been shown in Fig. 5.8. Table 5.1 shows different models of wavy fins used for their study. The correlations for j factor and friction factor have been developed by using 25 models and 250 simulated points, in both laminar and turbulent regimes.

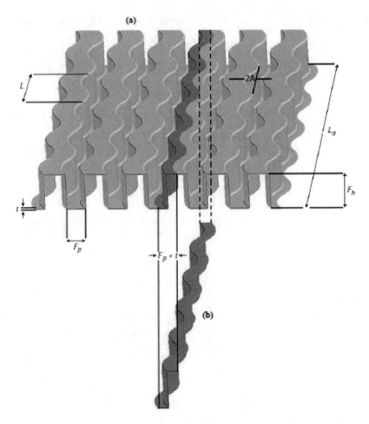

Fig. 5.8 Wavy plate-fin and its geometrical details (Khoshvaght Aliabadi et al. 2014)

Figure 5.9 shows the variation of *j* factor and *f* factor with Reynolds number for air (using correlations), water, and ethylene glycol (using simulation). They concluded that significant variation in *j* factor was observed for change of fluid. On the other hand, the change of working fluid had no effect on friction factor.

Wen et al. (2019) studied sine wavy fins for plate-fin heat exchanger and its optimization according to the fluid structure interaction analysis. They also studied the effect of variation in fin parameters on heat transfer and pressure drop characteristics. Increase in *j* factor with decreasing fin thickness and increasing fin space and fin height has been observed. Similar effect of fin thickness, fin spacing, and fin height on friction factor has been noted. The computational domain of a sinusoidal wavy fin for plate and fin heat exchangers has been shown in Fig. 5.10. The effect of double amplitude, fin thickness, wavelength, fin space, and fin height on JF factor has been shown in Figs. 5.11 and 5.12. The results for three optimal designs and the actual design have been tabulated in Table 5.2. They concluded that the performance of sine wavy fins with optimum design 1, optimum design 2, and optimum design 3 has been observed to be 11%, 8.4%, and 15.9% more than that of the original design in terms of JF factor.

Table 5.1 Different models of wavy fins used for the study (Khoshvaght Aliabadi et al. 2014)

Model no.	F_h	F_p	L	t	$2A$	L_d
1	10	3.5	10	0.3	1.0	50
2	8	1.5	9	0.5	1.0	72
3	10	2.5	8	0.1	2.5	64
4	9	1.5	10	0.2	2.5	70
5	10	2.0	7	0.5	2.0	49
6	10	3.0	9	0.2	0.5	81
7	7	3.5	7	0.2	1.5	56
8	7	3.0	11	0.1	1.0	77
9	6	1.5	7	0.1	0.5	35
10	9	2.5	7	0.4	1.0	63
11	7	2.5	10	0.5	0.5	60
12	6	2.5	9	0.3	1.5	63
13	8	3.5	8	0.4	0.5	56
14	9	3.5	9	0.1	2.0	54
15	6	2.0	8	0.2	1.0	48
16	9	2.0	11	0.3	0.5	88
17	6	3.5	11	0.5	2.5	99
18	10	1.5	11	0.4	1.5	66
19	8	2.5	11	0.2	2.0	55
20	8	2.0	10	0.1	1.5	90
21	6	3.0	10	0.4	2.0	80
22	7	2.0	9	0.4	2.5	45
23	7	1.5	8	0.3	2.0	72
24	9	3.0	8	0.5	1.5	40
25	8	3.0	7	0.3	2.5	42

Wen et al. (2016b) investigated the performance of serrated fins in PFHE. They used generic algorithm in order to obtain the optimal parameters of the serrated fin. The results obtained using optimal fin parameters have been presented in Table 5.3. They concluded that the performance of the heat exchanger was optimum when the fin having 9.5 mm height, 3 mm interrupted length, 2.6 mm fin spacing, and 0.1 mm thickness was used. The fin with optimal design showed 145 W increase in heat transfer rate and 0.117 W decrease in power consumption as compared to those of original serrated fin.

Wen et al. (2016a) have used multi-objective generic algorithm for performance optimization of PFHE having serrated fins. Their results of optimal performance of the heat exchanger have been shown in Table 5.4.

Muley et al. (2006) investigated the heat transfer augmentation characteristics of wavy plate-fin and have been shown in Fig. 5.13. Air was used as the working fluid for the analysis with $10 \leq Re \leq 1500$. Constant heat flux boundary condition was considered. The details of the wavy plate-fin channel configuration have been presented in Table 5.5. They defined "γ" to be the severity of waviness. They reported that for low Reynolds number ($Re < 100$), the flow in the tube with wavy

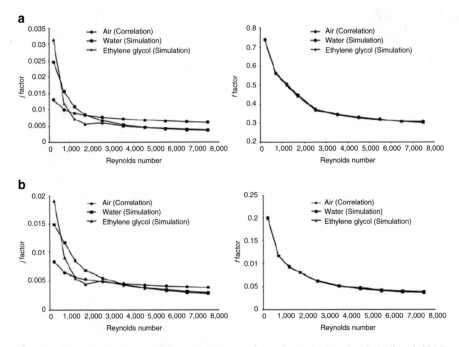

Fig. 5.9   Variation of j factor and f factor with Reynolds number (Khoshvaght Aliabadi et al. 2014)

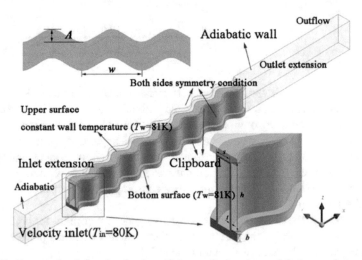

Fig. 5.10   Computational domain of a sinusoidal wavy fin for plate-and-fin heat exchangers (Wen et al. 2019)

plate-fin is similar to that in a smooth tube. In this region, the presence of wavy fin-plate aids in increased flow path and provides more time for heat transfer between the heated surface and the fluid. This results in increased j and f.

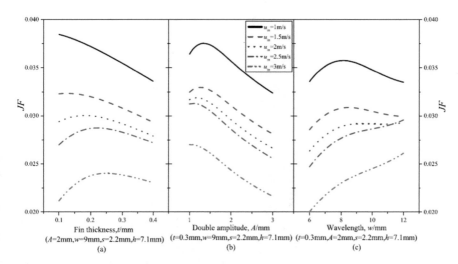

Fig. 5.11 Effect of (**a**) fin thickness, (**b**) double amplitude, and (**c**) wavelength on JF factor (Wen et al. 2019)

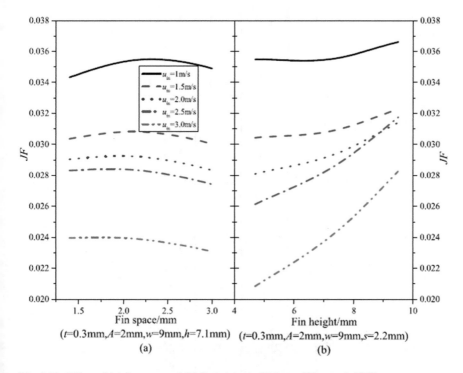

Fig. 5.12 Effect of (**a**) fin space and (**b**) fin height on JF factor (Wen et al. 2019)

Table 5.2 Results for three optimal designs and the actual design (Wen et al. 2019)

Items	Height/ mm	Space/ mm	Wavelength/ mm	Thickness/ mm	Double amplitude/mm	Inlet velocity/m/s	JF	Maximum stress/MPa
Original design	7.1	3	9	0.25	2	1	0.0347	21.7
Optimal design 1	8.89	1.45	6.78	0.15			0.0392	15.3
CFD validation error					2.03	1	0.0385	14.7
							1.8%	4.1%
Optimal design 2	8.86	1.48	6.78	0.14			0.0384	13.3
CFD validation error					1.44	1	0.0395	12.5
							2.8%	6.4%
Optimal design 3	8.86	1.48	7.28	0.19			0.0397	16.5
CFD validation error					1.80	1	0.0409	17.2
							2.9%	4.1%

Table 5.3 Results obtained using optimal fin parameters (Wen et al. 2016b)

	Height $h/$ mm	Space $s/$ mm	Thickness $t/$mm	Interrupted length $l/$mm	j	f	JF
Max j factor	9.48	2.72	0.11	3.18	**0.0195**	0.0623	0.0492
Validate					0.0190	0.0613	0.0482
Predicted error					2.6%	1.6%	2.1%
Min f factor	4.71	2.96	0.16	7.62	0.0122	**0.0383**	0.0362
Validate					0.0121	0.0396	0.0355
Predicted error					0.8%	−3.3%	2.0%
Max JF factor	9.45	2.62	0.11	3.01	0.0194	0.0606	**0.0494**
Validate					0.0191	0.0611	0.0485
Predicted error					1.6%	−0.8%	1.9%

Bold values indicate the optimum results by single objective optimization

Table 5.4 Results of optimal performance of the heat exchanger (Wen et al. 2016a)

Objectives	Re	h(mm)	s (mm)	t (mm)	l (mm)	j	f	JF
Max j factor	214	8.90	1.61	0.47	3.06	0.0393	0.2058	0.0665
Validate						0.0394	0.2045	0.0668
Predicted error						−0.2%	0.6%	0.5%
Min f factor	2971	4.87	1.62	0.11	8.78	0.0070	0.0233	0.0264
Validate						0.0073	0.0224	0.0260
Predicted error						−4.2%	4.1%	1.4%
Max JF factor	217	9.18	1.51	0.43	5.60	0.0358	0.1573	0.0668
Validate						0.0357	0.1543	0.0665
Predicted error						0.3%	1.9%	0.5%

The flow region for $Re > 100$ swirl flows results in increased momentum transport and heat transfer rate. This increase in heat transfer rate for both $Re < 100$ and $Re > 100$ is further augmented with increase in waviness severity of the wavy plate-fin, "γ." They concluded that the fin with 0.0667 waviness severity showed the superior performance in heat transfer enhancement based on area goodness factor. Moffat (1988) and Majumdar (2005) have also worked on heat transfer augmentation in wavy channels.

The numerical study on the heat transfer and pressure drop performance of PFHE using plain fins and serrated fins has been carried out by Wang et al. (2009). The PFHE with plain fins and serrated fins has been shown in Fig. 5.14a and b, respectively. They compared their numerical results with experimental results for plain and serrated fins which have been tabulated in Tables 5.6 and 5.7, respectively. Also, Table 5.8 shows the heat transfer rate and pumping power consumed by the

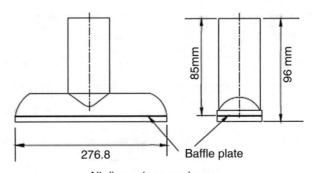

All dimensions are in mm

(a) Improvised header with a baffle

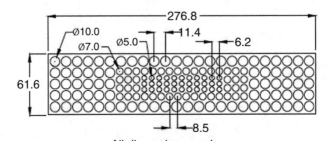

All dimensions are in mm

(b) Details of punched baffle plate

Fig. 5.13 (a) Header with a baffle and (b) baffle configuration (Sheik Ismail et al. 2009)

Table 5.5 Details of the wavy plate-fin channel configuration (Muley et al. 2006)

	Case 1	Case 2	Case 3
$\alpha = (S/H)$	0.1677	0.1677	0.1677
$\gamma = (2A/L)$	0.2666	0.1333	0.0667
$\varepsilon = (S/2A)$	0.803	0.803	0.803

Fig. 5.14 PFHE with plain fins (Wang et al. 2009)

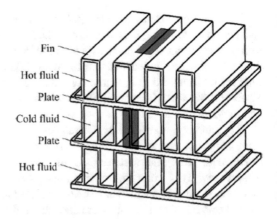

Table 5.6 Comparison of their numerical results with experimental results for plain fins (Wang et al. 2009)

Re	Experimental result				Numerical result		Error	
	$StPr^{2/3}$	f	Q, W	$\Delta p/L$, Pa/m	Q, W	$\Delta p/L$, Pa/m	Q, %	$\Delta p/L$, %
500	0.0103	0.0380	35.12	428.12	34.81	437.48	0.882	2.186
600	0.00890	0.0319	36.18	517.53	35.93	523.85	0.694	1.221
800	0.00704	0.0243	37.75	700.86	36.80	698.52	2.511	0.335
1000	0.00586	0.0198	38.98	892.30	37.34	883.80	4.217	2.074

Table 5.7 Comparison of their numerical results with experimental results for serrated fins (Wang et al. 2009)

Re	Experimental result				Numerical result		Error	
	$StPr^{2/3}$	f	Q, W	$\Delta p/L$, Pa/m	Q, W	$\Delta p/L$, Pa/m	Q, %	$\Delta p/L$, %
300	0.0295	0.131	5.93	1615.32	6.09	1626.07	2.724	0.666
400	0.0246	0.104	6.41	2279.81	6.41	2270.15	0.031	0.423
500	0.0216	0.088	6.86	3014.17	6.65	2942.26	3.071	2.386
600	0.0195	0.0772	7.27	3807.72	6.86	3629.4	5.613	4.683
800	0.0167	0.0645	8.00	5655.67	8.02	5243.87	0.258	7.281

Table 5.8 Heat transfer rate and pumping power consumed by the PFHE (Wang et al. 2009)

Re	Plain fin		Serrated fin	
	Q/F, kW/m^2	P/F, W/m^2	Q/F, kW/m^2	P/F, W/m^2
500	24.72	0.0549	67.37	0.370
600	25.52	0.0790	69.50	0.547
800	26.14	0.1404	81.25	1.054

PFHE using both the fins considered. They observed a twofold increase in heat flux using serrated fin over that of using plain fins. In order to study the effect of using fin on heat transfer rate and friction factor, the ratio of heat flux per pumping power consumption has been used which was noted to be 40% more for serrated fins as compared to that of the plain fin.

Ali and Ramadhyani (1992), Figs. 5.15 and 5.16, O'Brien and Sparrow (1982), Sparrow and Hossfeld (1984), and Molki and Yuen (1986) studied corrugated geometry for $150 < Re_{Dh} < 35{,}000$, made flow visualization, and observed that both j and f increase as the spacing increases.

Kays and London (1984) gave j and f versus Reynolds number curves for two wavy fin geometries and they observed competitive performance with that of the OSF (Rosenblad and Kullendorf 1975; Okada et al. 1972). They have used wavy channel geometries used in plate-type heat exchangers and obtained additional data for small-aspect-ratio channels.

3D corrugated channels are used in plate-type heat exchangers and rotary generator (Focke et al. 1985; Focke and Knibbe 1986; Stasiek et al. 1996; Abdel and Fletcher 1999, Fig. 5.17). Asymptotically, with the increase of corrugation angle, 3D

Fig. 5.15 PFHE with
serrated fins (Wang et al.
2009)

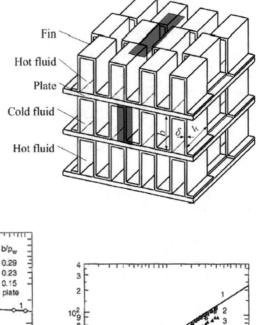

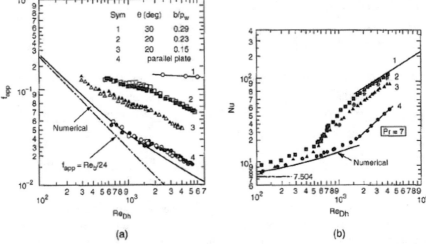

Fig. 5.16 Wavy channel data. (**a**) Friction factor, (**b**) Nusselt number for $Pr = 7$. Curve 1 $\theta = 30°$, $b/P_w = 0.29$, curve 2, $\theta = 20°$, $b/P_w = 0.23$, curve 3, $\theta = 30°$, $b/P_w = 0.15$, curve parallel plate (Ali and Ramadhyani 1992)

corrugation starts behaving like 2D corrugations and the predominant fluid flow along the furrows gets two sets of crisscrossing streams, which induce secondary swirling motions. First with smaller corrugation angle, the driving force producing the swirl in a furrow is the velocity component of the fluid moving along the opposite furrows in a direction perpendicular to the furrow.

Even lower corrugation angle made the interaction between fluid streams positive and each of the crossing streams had a velocity component in the same direction as the stream it crossed. For very high corrugation angle, the interaction is negative and cross streams had a retarding effect on each other. The flow pattern eventually changes and reflection between plate contact points makes the flow with zigzag

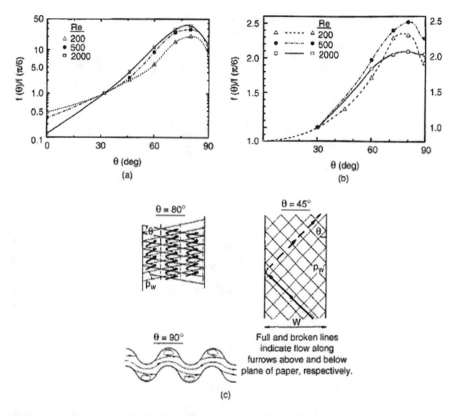

Fig. 5.17 The effect of the corrugated angle on the j and f factors for corrugated channel with $P_w/H = 2$: (**a**) friction factor, (**b**) j factor, (**c**) sketch of the flow pattern (Focke et al. 1985)

pattern. The change of flow pattern decreases j and f factors. Early transition to turbulent flow occurs.

Liquid crystal thermography (LCT) helps getting local heat transfer coefficients and the effect of corrugation pitch-to-height ratio (P_w/H) has been investigated by Stasiek et al. (1996).

Abdel and Fletcher (1999) correlations for j and f are as follows:

Laminar regime $Re_{Dh} < 500$:

$$Nu = 0.777 Re_{Dh}^{0.444} Pr^{0.4} \left(\frac{\theta}{45}\right)^{0.67} \tag{5.2}$$

$$f = 15 Re_{Dh}^{-0.3} \left(\frac{\theta}{45}\right)^{2.5} \tag{5.3}$$

Turbulent regime ($Re_{Dh} > 500$):

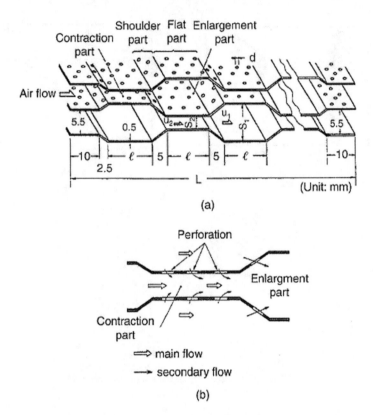

Fig. 5.18 (**a**) Perforated plate-fin geometry, (**b**) secondary flows through perforations (Fujii et al. 1988)

$$Nu = 0.26 Re_{\text{Dh}}{}^{0.67} Pr^{0.4} \left(\frac{\theta}{45}\right)^{0.67} \tag{5.4}$$

$$f = 7.3 Re_{\text{Dh}}{}^{-0.198} \left(\frac{\theta}{45}\right)^{2.5} \tag{5.5}$$

If the porosity is sufficiently high, boundary layer dissipates in the wake region formed by the holes or slots. This results in enhancement. This type of enhancement is more likely for transient and turbulent flow (Shah 1975; Shen et al. 1987). However, the performance of OSF is much better than the perforated fins and in perforated fins there is much wastage of materials because of its removal from the surface. Secondary flow through the perforation is an important factor for enhancement, Figs. 5.18 and 5.19 (Fujii et al. 1988, 1991; Xuan et al. 2001). Xuan et al. (2001) have compared their data with those of the wavy surface given by Kays and London (1984).

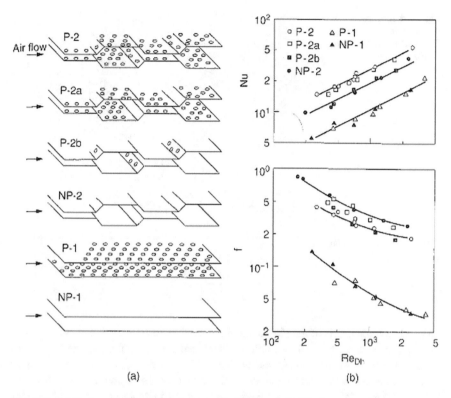

Fig. 5.19 (**a**) Perforated plate geometries tested by Fujii et al. (1988), (**b**) test results on illustrated surfaces

Choudhury and Garg (1991), Focke and Knibbe (1986), Focke et al. (1985), Gaiser and Kottke (1990), Heggs and Walton (1999), Hessami (2002), Hotani et al. (1977), Metwally and Manglik (2004), Oosthuizen and Garrett (2001), and Ros et al. (1995) investigated the performance of corrugated fins.

A special weave of screen wire is formed by modifying a pin-fin surface geometry. The wire may have a round, elliptical, or square cross-section shape. Even though there may be benefits of enhancement by pin fin, it still is very costly compared to OSF and louver geometries. Information on pin-fin geometries may be obtained from Theoclitus (1966), Hamaguchi et al. (1983), Torikoshi and Kawabata (1989), and Ebisu (1999). Boundary layers form on the wires and these get dissipated in the void region between the wire attachment points. The matrix material may be corrugated to form the packing of a plate-and-fin heat exchanger. The layers of the expanded metal may be compressed to conform to the tube shape and soldering it to parallel tubes, Fig. 5.20.

Matsumoto et al. (2000) studied the heat transfer characteristics of end wall with a single oblique pin fin. Pin fin was inclined to the streamwise direction at every 15° in the range of $\theta = -60°$ (upstreamwise) to $\theta = +60°$ (downstreamwise), where θ is an

Fig. 5.20 Expanded metal (copper) matrix geometry (Torikoshi and Kawabata 1989)

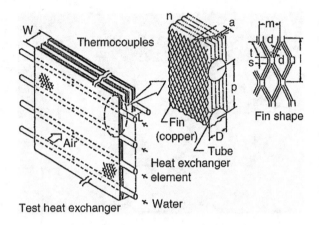

Table 5.9 Experimental conditions (Matsumoto et al. 2000)

U_{m} (m/s)l	2.9	4.3	5.7	7.2
Re	3.7×10^3	5.4×10^3	7.3×10^3	9.1×10^3
Re_{H}	1.8×10^4	2.7×10^4	3.6×10^4	4.5×10^4

angle between the axis of the pin fin and the normal of the end wall. The heat transfer coefficient on the end wall and area of enhanced heat transfer became maximum when pin-fin inclined upstreamwise at $\theta = -45°$. They observed that heat transfer can be enhanced by the impingement of wake flow on downstream end wall of the pin fin at $\theta = -45°$. It was also seen from the results that there was no heat transfer enhancement on the end wall in case of pin-fin inclined downstreamwise.

Many studies on the heat transfer enhancement of the end wall with short pin fins were carried out by Saboya and Sparrow (1976), Lau et al. (1987), Chyu and Goldstein (1991), and Matsumoto et al. (2000). Two types of Reynolds number were taken: Re based on the pin-fin diameter and Re_{H} based on the hydraulic diameter of the channel. Table 5.9 shows the experimental conditions. Figure 5.21 shows the comparison of Nusselt number of oblique pin fin to without pin fin for different values of Reynolds number. The relation between average Nusselt number and Reynolds numbers is defined as

$$Nu_{\mathrm{AUG}} = a \times Re^b \qquad (5.6)$$

The values of coefficient (a) and exponent (b) are listed in Table 5.10.

Hwang et al. (2000) studied the transient solution of 2D cylindrical pin fin with constant heat flux condition. The tip of pin fin was subjected under convective effect. The analytical transient solution had been determined by using the Laplace transformation and the separation of variables method. Yang (1972), Aziz (1975), Suryanarayana (1975), and Mao and Rooke (1994) investigated the transient heat transfer problem of one-dimensional fins.

Fig. 5.21 Relationship between averaged Nusselt number and Reynolds number (Matsumoto et al. 2000)

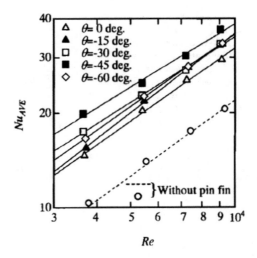

Table 5.10 Shows values of coefficient a and exponent b (Matsumoto et al. 2000)

Oblique angle θ (°)	a	b
−60	0.0312	0.765
−45	0.0798	0.671
−30	0.0593	0.693
−15	0.0158	0.839
0	0.0269	0.769

Fig. 5.22 Geometry of 2D cylindrical pin fin (Hwang et al. 2000)

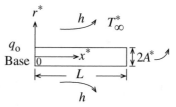

Chu et al. (1983) studied the transient response of circular pin fins under variable heat flux. Figure 5.22 shows the geometry of 2D cylindrical pin fin. They defined the ratio of the transient heat flow rate on the tip surface to the transient total heat flow rate on the fin surface for quantifying the fin tip convective effects:

$$R_{QT}(\%) = \frac{Q_L(t)}{Q(t)} \times 100\% \qquad (5.7)$$

Figure 5.23a and b illustrates the variation of total heat flow rate with respect to time under the conditions of various values of geometric parameter (G), and ratio of convective heat transfer coefficient (H) at two different transversal Biot number (Bi_a). Figure 5.24a and b shows the effect of H, G, and Bi_a on the ratio R_{QT} with respect to time.

Fig. 5.23 Effects of the geometry parameter G and the ratio of convective heat transfer coefficient H on the transient total heat flow rate on the surface in 2D pin fin. (a) $Bi_a = 0.01$, (b) $Bi_a = 1$ (Hwang et al. 2000)

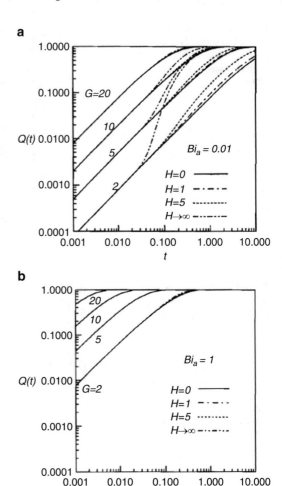

Saha (2008) investigated the unsteady flow and heat transfer conducted in three-dimensional periodic array of cubic pin fins located inside a channel. The numerical investigation was carried out at a particular Reynolds number 7125, for pin fin placed in an in-line pattern inserted in channel. The pattern is periodic in two dimensions both streamwise and transversewise. The periodicity of pattern was 2.5 times the pin-fin dimension. He solved Navier-Stokes and energy equations taking help of higher order temporal and spatial discretization methods. He observed shorter shear layer in square pin-fin geometry.

There was no significant flow periodicity in transverse as well as in same plane for vorticity contours. He observed that flow periodicity and geometric periodicity were same based on time-averaged flow field. He drew some conclusions on numerical analysis that instantaneous flow and temperature fields were unaffected by the flow periodicity in the domain of selected inter-rib modules. Also, observation was that

Fig. 5.24 Effect of the geometry parameter G and the ratio of heat transfer coefficient H on the ratio R_{QT} (%). (**a**) $Bi_a = 0.01$, (**b**) $Bi_a = 1$ (Hwang et al. 2000)

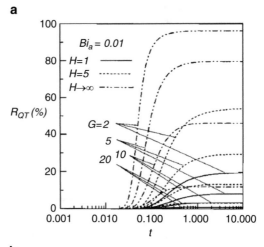

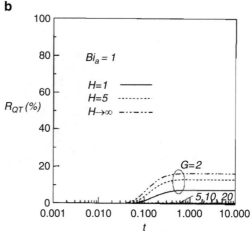

the streamwise and transversewise periodic lengths make variation to the rms fluctuations.

Li and Chen (2005) investigated numerically and experimentally the pin-fin geometries attached to the heat sinks with air impingement cooling. Maveety and Jung (2002) and Kim and Kuznetsov (2003) investigated the same pin-fin heat sinks. Li and Chen (2005) presented Table 5.11 for experimental dimensions of heat sinks. They show the simulation domains and boundary conditions in Fig. 5.25. They numerically and experimentally claimed that results were within relative error of 6%. They have observed velocity and temperature fields and concluded that the experimental results and simulation data of surface temperature distributions are based on the heat sinks at various positions. Figure 5.26 presents the path lines and corresponding speed of channel. They found that increased Reynolds number from 5000 generated more circulations on the cross section.

Table 5.11 Dimensions of heat sink (Li and Chen 2005)

No.	W (mm)	H (mm)	A (mm^2)
1	6.5	35	39,160
2		40	43,840
3		45	48,520
4	8.0	35	46,720
5		40	52,480
6		45	58,240
7	9.5	35	54,280
8		40	61,120
9		45	67,960

$L = 80$ mm, $b = 8$ mm

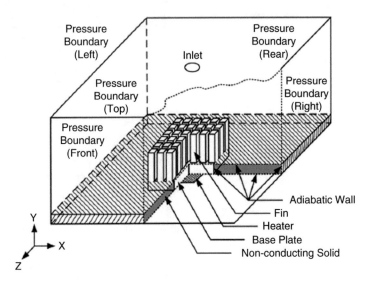

Fig. 5.25 Computational domain and boundary conditions (Li and Chen 2005)

Another aspect put forwarded in Figs. 5.27 and 5.28 is that the path lines and corresponding speed on section intersecting fins are involved. They studied that the temperature distribution of heat sink diminishes moving from bottom to top and it increases from inner to outer direction. They studied the fin width influence on thermal resistance and observed that it lowered down as fin width increased. These results have been compared with simulated data in Table 5.12. Similarly Table 5.13 has been presented for evaluation of fin height impact on thermal resistance. They concluded that for fin height up to 45 mm there was decrease in the thermal resistance whereas beyond 45 mm the decreased performance of heat sink has been observed.

Park et al. (2007) worked on maximizing the heat transfer by considering pin-fin-type heat sinks solutions. They modelled rectangular and circular pin fin made up of aluminum. The objective of the study is to optimize heat sink shape, the thermal

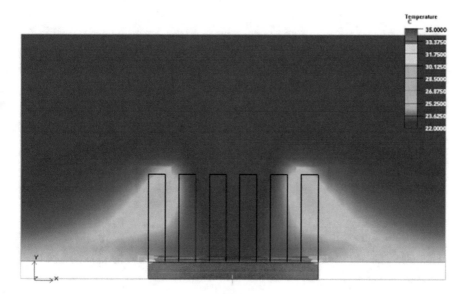

Fig. 5.26 Temperature distribution on section Channel 1 (Li and Chen 2005)

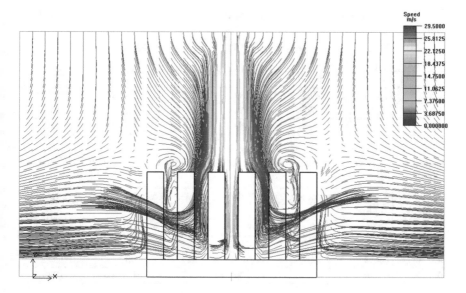

Fig. 5.27 Pathlines and corresponding speeds on section Fin 1 (Li and Chen 2005)

resistance and the pressure drop occurred. For this problem, they considered steady and incompressible fluid. The fluid flow was three-dimensional turbulent mixed convective flows with all properties constant except density. The symmetry eases the computation as one-quarter physical domain is sufficient. The initial volume of both

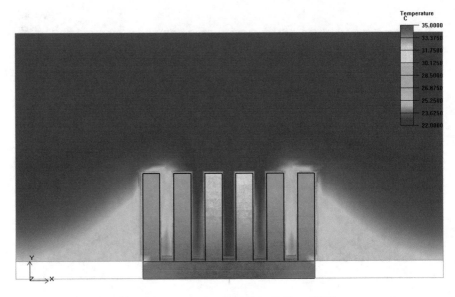

Fig. 5.28 Temperature distribution on section Fin 1 (Li and Chen 2005)

Table 5.12 Effects of fin width and fin height on thermal resistance at $Re = 5000$ (Li and Chen 2005)

W (mm)	Simulated thermal resistance (°C/W)			Measured thermal resistance (°C/W)		
	H (mm)					
	35	40	45	35	40	45
6.5	1.033	0.932	0.859	0.991	0.968	0.929
8.0	0.826	0.763	0.704	0.904	0.933	0.883
9.5	0.728	0.690	0.637	0.840	0.883	0.864

Table 5.13 Effects of fin width and fin height on thermal resistance at $Re = 25,000$ (Li and Chen 2005)

W (mm)	Simulated thermal resistance (°C/W)			Measured thermal resistance (°C/W)		
	H (mm)					
	35	40	45	35	40	45
6.5	0.517	0.488	0.455	0.447	0.400	0.379
8.0	0.433	0.408	0.385	0.377	0.378	0.376
9.5	0.348	0.329	0.309	0.356	0.337	0.339

fins was made same and the design variables with their objective function have been presented in Table 5.14. The optimized values of circular and rectangular fins are featured in Table 5.15 as the maximum temperature of 336 K, pressure drop variation with increasing T_{max}, design variable configurations, different maximum temperature relation with pressure drop, volume, fin diameter, etc. However, it should be noted that pressure drop is inversely proportional to maximum

Table 5.14 Initial value of design variables and their objective functions (Park et al. 2007)

		Circular fin	Rectangular fin
Design variables	D	3.38 mm	3.00 mm
	h	45.0 mm	45.0 mm
	c	3.0 mm	3.0 mm
Pressure drop	ΔP	3.63 Pa	3.95 Pa
Maximum temperature	ΔT	339.9 K	338.8 K

Table 5.15 Optimal values of the design variables for circular and rectangular fins at $T_{\max} < 336$ K (Park et al. 2007)

		Circular fin	Rectangular fin
Design variables	D	4.42 mm	3.59 mm
	h	48.1 mm	49.5 mm
	c	3.64 mm	3.67 mm
Pressure drop	ΔP	5.52 Pa	5.11 Pa
Thermal resistance	θ_j	0.575 K/W	0.575 K/W
Volume	V	107.5 cm^3	96.6 cm^3

temperature. The pressure drop determines the cost of the specific device as it dictates the size of the required fan. They used CFD code and mathematical optimization techniques and concluded that for optimum model at maximum temperature 336 K, pressure drop gain is of 29.4% and thermal resistance fell down by 14.8% in comparison to initial model. They found that the optimum design variables were $D = 3.59$ mm, $h = 49.54$ mm, and $c = 3.64$ mm. The results supported rectangular fin-attached heat sinks than that of circular fin.

Donahoo et al. (2001) investigated the cross-pin configuration designed for turbine blade. This type of configuration is provided to enhance the cooling rate of blade. The extended surface of the cross pin provided structural integrity and stiffness to the blade itself. They presented two-dimensional numerical simulation of coolant airflow through turbine blade cooling passage. Circular pins were arranged in staggered manner with varying the pin space. They simulated the model with specified parameter such as viscous flow, wide range of Reynolds numbers, and optimum spacing.

They optimized the specified parameters for which maximum heat transfer and minimum total pressure drop occurred. Pareto plots identified the optimum data points graphically. The results showed that cross-pin heat transfer increased up to certain number of rows, and then decreased gradually in subsequent rows. Table 5.16 shows the total pressure drop, ratio of outlet to inlet temperatures, and min-max calculations for heat flux in different cases. Figure 5.29 shows the generic Pareto curve between two objective functions. The more work has been done on cross-pin fin by Peng (1983), Armstrong and Winstanley (1988), Metzger et al. (1982, 1984, 1986), Van Fossen (1981), and Chyu (1990, 1998).

Minakami et al. (1995) conducted an experiment by using wind tunnel to find the distribution of local heat transfer coefficient in a pin-fin array. Heat transfer performance was investigated by local heating method. Electric current was used to heat the pin rows. The flow pattern in the pin arrays and effect of an oscillating flow were studied experimentally. Fin rows were heated in two ways: heated pin rows

Table 5.16 Passage flow test cases and results where $\text{Max}\Delta f = \max\left(f_i\left(\dot{x}\right) - f_i/f_i^*\right)$

		Set A				Set B		
		$Re_D = 1270$				$Re_D = 3980$		
		$M_i = 0.024$				$M_i = 0.076$		
		$U_i = 15.48$ m/s				$U_i = 48.51$ m/s		
		$T_i = 1000$ K				$T_i = 1000$ K		
		$T_s = 1300$ K				$T_s = 1300$ K		
x/D	q'' (W/cm²)	T_o/T_i	ΔP_t (kPa)	Max Δf	q'' (W/cm²)	T_o/T_i	ΔP_t (kPa)	Max Δf
1.00	2.74	1.74	1.70	1.48	6.70	1.42	18.22	1.77
1.10	3.00	1.73	1.57	1.37	5.84	1.52	12.80	0.81
1.25	3.03	1.68	1.05	0.59	6.34	1.50	10.24	0.51
1.50	2.80	1.62	0.83	0.18	6.22	1.52	8.30	0.19
1.75	2.58	1.62	0.81	0.13	5.85	1.53	7.74	0.05
2.00	2.67	1.60	0.71	0.04*	5.34	1.53	7.90	0.19
2.25	2.55	1.60	0.71	0.08	5.62	1.56	6.84	0.01*
2.50	2.59	1.60	0.66	0.14	5.18	1.54	7.88	0.03
2.75	2.44	1.61	0.71	0.12	5.26	1.53	7.13	0.13
3.00	2.43	1.60	0.67	0.18	5.06	1.54	7.58	0.10
3.25	2.31	1.64	0.70	0.18	5.08	1.53	6.96	0.18
3.50	2.27	1.60	0.68	0.22	5.12	1.56	6.58	0.24
3.75	2.25	1.62	0.70	0.20	4.95	1.54	6.95	0.20
4.00	2.21	1.63	0.69	0.23	4.86	1.54	6.95	0.22
		Set C				Set D		
		$Re_D = 7310$				$Re_D = 13{,}800$		
		$M_i = 0.140$				$M_i = 0.265$		
		$U_i = 89.10$ m/s				$U_i = 168.20$ m/s		
		$T_i = 1000$ K				$T_i = 1000$ K		
		$T_s = 1300$ K				$T_s = 1300$ K		
x/D	q'' (W/cm²)	T_o/T_i	ΔP_t (kPa)	Max Δf	q'' (W/cm²)	T_o/T_i	ΔP_t (kPa)	Max Δf
1.00	11.0	1.48	58.46	2.02	18.81	1.52	200.45	2.28
1.10	9.03	1.48	39.52	0.86	16.14	1.61	132.49	1.03
1.25	9.67	1.56	31.00	0.48	15.94	1.63	102.90	0.53
1.50	9.18	1.61	25.55	0.16	15.05	1.69	83.68	0.17
1.75	8.75	1.63	23.37	0.01*	14.21	1.70	76.30	001*
2.00	8.37	1.62	24.31	0.02	14.07	1.68	79.60	0.05
2.25	8.31	1.67	20.22	0.02	13.53	1.75	65.77	0.20
2.50	8.04	1.62	24.17	0.02	13.45	1.68	80.44	0.03
2.75	7.95	1.62	20.94	0.20	13.09	1.72	67.14	0.21
3.00	7.87	1.63	25.55	0.04	13.08	1.71	75.32	0.07
3.25	7.58	1.63	23.36	0.10	12.30	1.72	67.05	0.25
3.50	7.62	1.67	19.34	0.31	11.94	1.75	61.14	0.36
3.75	7.37	1.64	20.68	0.26	11.88	1.73	65.96	0.29
4.00	7.24	1.65	20.58	0.28	11.73	1.73	66.46	0.29

(*) denotes lowest values corresponding to optimum spacing (Donahoo et al. 2001)

Fig. 5.29 Generic Pareto curve (Donahoo et al. 2001)

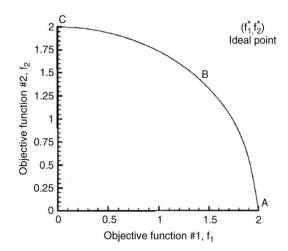

individually and heated all pin rows simultaneously under uniform heat flux. Distribution of local heat transfer coefficient of individual heating and simultaneous heating for $Re = 20$, $Re = 40$, and $Re = 60$ are as shown in Figs. 5.30, 5.31, and 5.32, respectively. They observed that local heat transfer coefficient was constant for individual heating pin fin and decreased for simultaneous heating pin fin at low Reynolds number. Mochizuki and Yagi (1990), Minakami and Mochizuki (1992), and Minakami et al. (1993a, b) studied the performance evaluation of heat transfer characteristics on pin-fin heat sink.

Aihara et al. (1990) and Park et al. (2004) studied the performance of pin-fin arrays in plate-fin heat exchangers. Alkam et al. (2001) and Chen and Chen (1990) worked on porous fins. Anashkin et al. (1976), Hendricks et al. (1995), Kovalenko et al. (1980), Mikulln et al. (1979), Nilles et al. (1995), Riddell (1966), Rodriquez and Mills (1996), Souidi and Bontemps (2001), and Xuan et al. (2001) have presented the performance of perforated fins.

Metal foam fin geometry for enhancement has been tested by Kim et al. (2000). Bhattacharya and Mahajan (2000) and Klett et al. (2000) used metal foams for electronic heat sinks. The pore density (PPI) and porosity (ε) are important for actual heat transfer enhancement, Figs. 5.33, 5.34, 5.35, and 5.36. The enhancement in metal foam fin is due to high surface-area-to-volume ratio and enhanced flow mixing due to the tortuous passage. But very small hydraulic diameter causes huge pressure drop which is much larger compared to other enhanced fin geometries. Klett et al. (2000) worked with high thermal conductivity being higher than aluminum foam.

Information on the use of very common plain fins may be obtained from Shah and London (1978), Kays and London (1984), and Tischenko and Bondarenko (1983). For plain fins, the thermal entrance length effect is very important for the flow that may be developing rather than being fully developed, Figs. 5.37 and 5.38. However, for interrupted fin heat exchangers, the entrance effect is annulled due to the periodic flow interpretation. Rotary or valued regenerators for combustors transfer heat through small hydraulic diameter packing. The design, however, may have full-

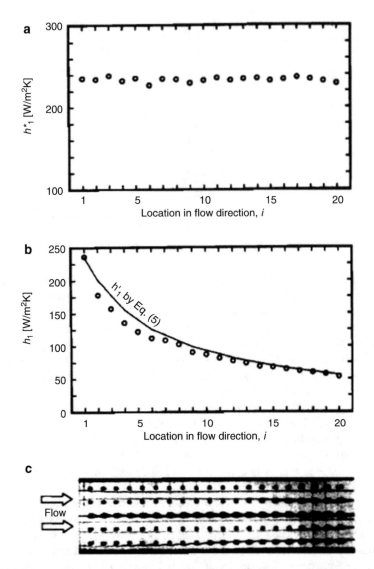

Fig. 5.30 Test results for $Re = 20$ (Minakami et al. 1995). (a) Distribution of local heat transfer coefficient in case of heating individual pin rows separately. (b) Distribution of local heat transfer coefficient in case of heating all pin rows simultaneously in a uniform heat flux. (c) Visualized flow pattern

height, continuous fins aligned with the flow at discrete spacing. The valve-type glass furnace regenerator has "brick checkers" geometry. Notched plate packings and 3D corrugated channels are also used in rotary regenerators, Fig. 5.39.

Sparrow et al. (1977), Patankar and Prakash (1981), Kelkar and Patankar (1989), Weiting (1975), Suzuki et al. (1994), Xi et al. (1995), Zhang et al. (1997), Mercier

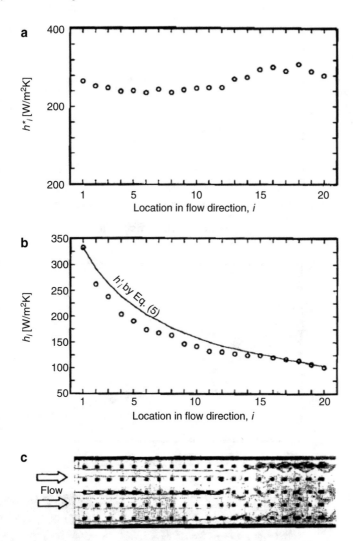

Fig. 5.31 Test results for $Re = 40$ (Minakami et al. 1995). (**a**) Distribution of local heat transfer coefficient in case of heating individual pin rows separately. (**b**) Distribution of local heat transfer coefficient in case of heating all pin rows simultaneously in a uniform heat flux. (**c**) Visualized flow pattern

and Tochon (1997), Xi and Shah (1999), Mochizuki et al. (1988), London and Shah (1968), Kajino and Hiramatsu (1987), Achaichia and Cowell (1988), Webb and Trauger (1991), Suga and Aoki (1991), Atkinson et al. (1998), Tafti et al. (2000), DeJong and Jacobi (2003), Heikal et al. (1999), Asako and Faghri (1987), Yang et al. (1995), Ergin et al. (1997), Kouidry (1997), McNab et al. (1998), Abou-Madi (1998), Xu and Min (2004), Ali and Ramadhyani (1992), Cicfalo et al. (1996), and Sunden (1999) may be referred for numerical simulation modelling and analysis

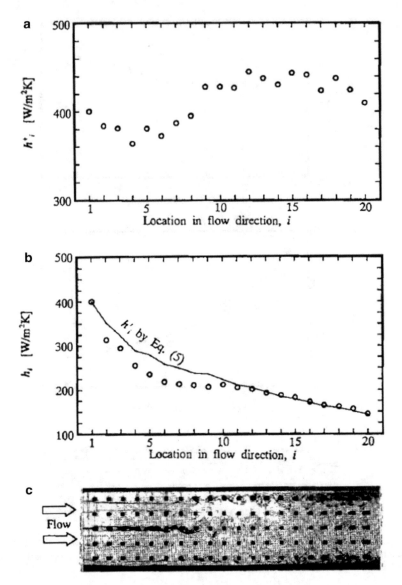

Fig. 5.32 Test results for $Re = 60$ (Minakami et al. 1995). (**a**) Distribution of local heat transfer coefficient in case of heating individual pin rows separately. (**b**) Distribution of local heat transfer coefficient in case of heating all pin rows simultaneously in a uniform heat flux. (**c**) Visualized flow pattern

and comparison of results with experimental data of offset-strip fins, louver fins, wavy channels, chevron plates, and other plate-and-fin extended surfaces.

Various numerical techniques like SIMPLE (and its improved versions), vorticity-based methods, different turbulence modelling, and simulations like LES

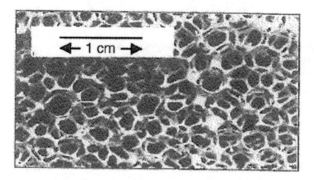

Fig. 5.33 Photo of a metal foam (5PPI) (Bhattacharya and Mahajan 2000)

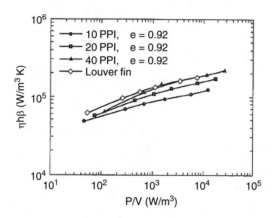

Fig. 5.34 The air-side performance per unit volume plotted as a function of power consumption per unit volume (P/V) for porous fin ($\varepsilon = 0.2$) and a louver fin ($L_P = 1.0$ mm, $P_f = 1.88$ mm) (Kim et al. 2000)

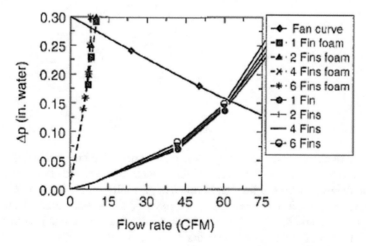

Fig. 5.35 Pressure drop vs. flow rate curves for various heat sink having longitudinal fins and fins with fin gap filled with 20 PPI metal foam, along with the fan curve (Bhattacharya and Mahajan 2000)

Fig. 5.36 Comparison of fin efficiency of three mesh materials

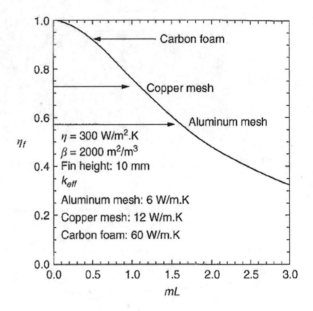

Fig. 5.37 Nu_m/Nu_{fd} for laminar entrance region flow with developed velocity profile in different channel shapes for constant wall for constant wall temperature boundary condition (Webb 1987)

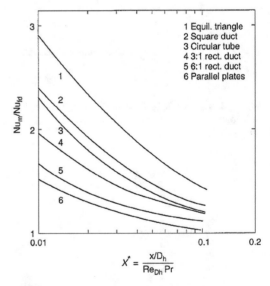

and DNS methods have been employed and modern-day computational process using very powerful digital supercomputers has been tried. However, very often underprediction, overprediction, and limited applicability of the data, methods, and techniques employed have been the experiences of the investigations. Analysis of the flow pattern and heat transfer near the fin surface has shown that heat transfer is enhanced by local instabilities near the surface, which are created by upstream vortices that impinge on the fin.

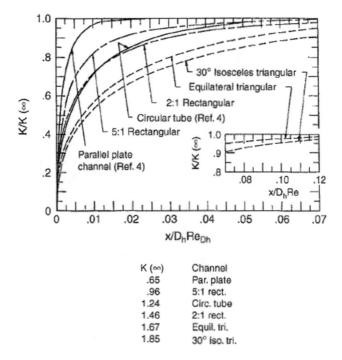

K (∞)	Channel
.65	Par. plate
.96	5:1 rect.
1.24	Circ. tube
1.46	2:1 rect.
1.67	Equil. tri.
1.85	30° iso. tri.

Fig. 5.38 $K(x)/K(\infty)$ for laminar flow in different duct shapes (Webb 1987)

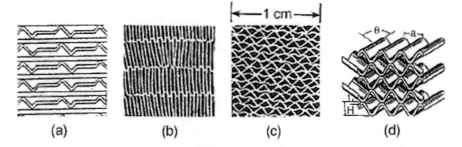

Fig. 5.39 Matrix geometries for rotary regenerators: (a) notched plate matrix, (b) deep-fold rectangular matrix, (c) triangular matrix, (d) corrugated channels (Cicfalo et al. 1996)

Steady flow calculations often fail to capture enhanced large-scale mixing in the unsteady laminar flow regime. For high Reynolds number flow, intrinsic 3D effects become important. Sometimes, numerical predictions do tally with the flow visualization. But this is not universally true. This discrepancy is perhaps due to inherent uncertainties, instabilities, and other associated difficulties and limitations in applying, for example, a periodic boundary condition for a fully developed solution.

Yang et al. (2016) numerically studied that a wing panel with perforations be installed in a header of a plate-fin heat exchanger in order to improve its thermal

Fig. 5.40 Geometrical details of perforated wing panel (Yang et al. 2016)

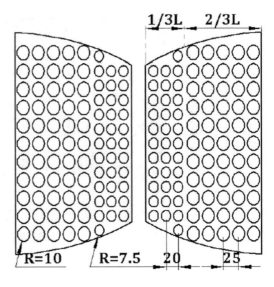

Fig. 5.41 Optimal design parameters of the header (Yang et al. 2016)

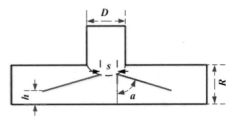

effectiveness. In convectional header configuration, a drop in velocity of the fluid flow at the exit of the header is noted due to maldistribution of the fluid flow, which eventually degrades the effectiveness of the heat exchanger. This problem has been overcome by the use of perforated wing panels. The geometrical details of the perforated wing panel have been shown in Fig. 5.40. The optimization design parameters of the header have been shown in Fig. 5.41.

The effect of geometrical parameters of the wing panel such as wing relative height, wing space ratio, and wing angle on performance parameters, namely, effectiveness (ε), flow maldistribution parameter (S_v), effectiveness degradation rate ($\Delta\varepsilon$), and power consumption (P), has been presented. The variation in flow maldistribution parameter, power consumption, and effectiveness degradation rate with Reynolds number has been shown in Figs. 5.42, 5.43, and 5.44, respectively. The graphs have been plotted for both the conventional header configuration and the modified header configuration with perforated wing panels for comparison. They reported that the optimum performance was obtained corresponding to wing relative height of 0.1, wing space ratio of 0.1, and wing angle of 70. They concluded that the decrease in flow maldistribution parameter and effectiveness degradation rate and increase in pumping power penalty by using the improved header configuration were 62.2–65.1%, 91.9–93%, and 88.1–90%, respectively.

Fig. 5.42 Flow
maldistribution parameter
S_v vs. Reynolds number
(Yang et al. 2016)

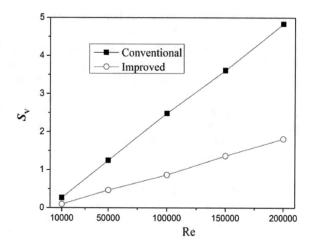

Fig. 5.43 Power
consumption vs. Reynolds
number (Yang et al. 2016)

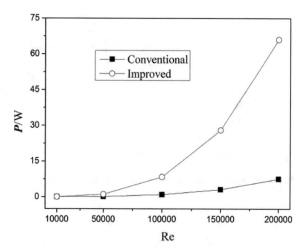

Fig. 5.44 Effectiveness
degradation
rate vs. Reynolds number
(Yang et al. 2016)

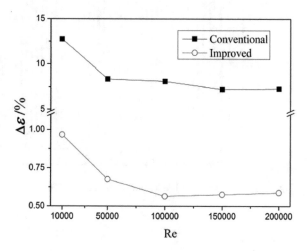

Wen et al. (2006) carried out both experimental and numerical study of flow patterns at the inlet of plate-fin heat exchangers. They have proposed a modified header configuration having punched baffles, in order to reduce the entrance flow maldistribution. The modified header design has been shown in Figs. 5.45 and 5.46. Baffle positioning in the header and baffle construction has been shown in Fig. 5.47.

Wen et al. (2016a, b, c), Kim and Byun (2012), Kim and Kim (2010), Wang et al. (2009), and Singh et al. (2014) studied the effect of nonuniformity of the flow on the performance of plate-fin heat exchangers. Baek et al. (2014), Jung et al. (2012), Wen et al. (2006), and Liu et al. (2007) proposed various inlet configurations for the plate-fin heat exchangers in order to achieve uniformity of the flow.

Cuta (1998), Iwasaki et al. (1994), Kawamura et al. (2001), Lu et al. (2004), Shwaish et al. (2001, 2002), Iyengar and Bar-Cohen et al. (2002), Sasao et al. (1999), Soodphakdee et al. (2000), Yu et al. (2005), Kelkar and Patankar (1987), Kirpikov and El'chinov (1988), Kreid et al. (1979), Lee et al. (2001), Lee (1986), Li et al.

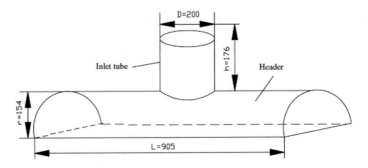

Fig. 5.45 Modified header configuration (Wen et al. 2006)

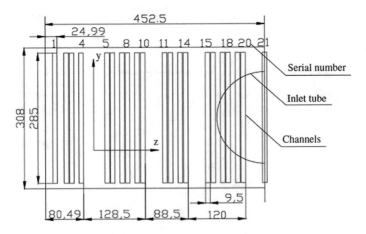

Fig. 5.46 Channels at the header outlet (Wen et al. 2006)

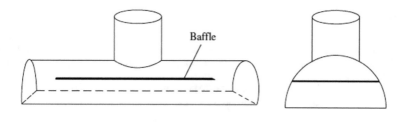

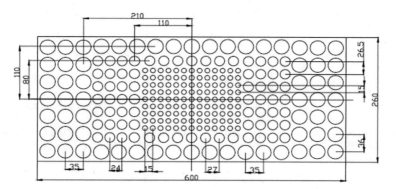

Fig. 5.47 Baffle positioning in the header and baffle construction (Wen et al. 2006)

(1999), Lin and Lee (1998, 2000), Loehrke et al. (1979), Michallon et al. (1996), Mullisen and Loehrke (1983), Naik et al. (1986), Nunez and Polley (1999), Onishi et al. (2001), Picon-Nuñez et al. (1999), Prakash and Lounsbury (1986), Prakash and Sparrow (1980), Rocha et al. (1997), Rodrigues and Yanagihara (1999), Rosman et al. (1984), Saha and Acharya (2004), Shohtani (1990), Thomas (1966), Tischenko and Bondarenko (1983), Tishchenko et al. (1979), Tolpadi and Kuehn (1984), Yakushin (1977), Ahmed et al. (2012), Althaher et al. (2012), Aris et al. (2011), Chen & Shu (2004), Chomdee & Kiatsiriroat (2006), Dake & Majdalani (2009), Ferrouillat et al. (2006), Henze & von Wolfersdorf (2011), Henze et al. (2011), Holmes (1995), Huisseune et al. (2013a), Huisseune et al. (2013b), Joardar & Jacobi (2007), Leu et al. (2004), Li et al. (2013), Li et al. (2017), Maiti (2002), Manson (1950), Min et al. (2010), Minakami et al. (1992), Mochizuki & Yagi (1975), Pesteei et al. (2005), Ranganayakulu et al. (2006), Sinha et al. (2013), Song et al. (2017), Suga & Aoki (1995), Suga et al. (1990), Tian et al. (2009), Toyoshima et al. (1986), Yang et al. (2010), Zhang & Yanzhong (2003), Wieting (1975) and Zhang et al. (1997) are the different researchers who considered various methods to improve the heat transfer augmentation in plate-fin heat exchangers.

References

Abdel-Kariem AH, Fletcher LS (1999) Comparative analysis of heat transfer and pressure drop in plate heat exchangers. In: Proc of the 5th ASME/JSME thermal eng conf, San Diego, CA, AJTE99-6291

Abed WM, Whalley RD, Dennis DJ, Poole RJ (2015) Numerical and experimental investigation of heat transfer and fluid flow characteristics in a micro-scale serpentine channel. Int J Heat Mass Transfer 88:790–802

Abou-Madi M (1998) A computer model for mobile air-conditioning system. PhD thesis, University of Brighton, Brighton, UK

Achaichia A, Cowell IA (1988) Heat transfer and pressure drop characteristics of flat tube and louvered plate fin surfaces. Exp Therm Fluid Sci 1:147–157

Ahmed HE, Mohammed HA, Yusoff MZ (2012) An overview on heat transfer augmentation using vortex generators and nanofluids: approaches and applications. Renew Sust Energ Rev 16:5951–5993

Aihara T, Maruyama S, Kobayakawa S (1990) Free convective/radiative heat transfer from pin-fin arrays with a vertical base plate (general representation of heat transfer performance). Int J Heat Mass Transfer 33(6):1223–1232

Ali MM, Ramadhyani S (1992) Experiments on convective heat transfer in corrugated channels. Exp Heat Transfer 5:175–193

Alkam MK, Al-Nimr MA, Hamdan MO (2001) Enhancing heat transfer in parallel-plate channels by using porous inserts. Int J Heat Mass Transfer 44:931–938

Althaher MA, Abdul-Rassol AA, Ahmed HE, Mohammed HA (2012) Turbulent heat transfer enhancement in a triangular duct using delta-winglet vortex generators. Heat Transfer Asian Res 41:43–62

Armstrong JE, Winstanley DA (1988, Jan 1) A review of staggered array pin fin heat transfer for turbine cooling applications. J Turbomach 110(1):94–103

Anashkin OP, Keilin VE, Patrikeev VM (1976) Compact high efficiency perforated-plate heat exchangers(in helium refrigerators). Cryogenics 16:437–439

Aris MS, McGlen R, Owen I, Sutcliffe CJ (2011) An experimental investigation into the deployment of 3-D, finned wing and shape memory alloy vortex generators in a forced air convection heat pipe fin stack. Appl Therm Eng 31:2230–2240

Asako Y, Faghri M (1987) Finite volume solutions for laminar flow and heat transfer in a corrugated duct. J Heat Transfer 109:627–634

Atkinson KN, Drakulic R, Heikal MR, Cowell IA (1998) Two and three-dimensional numerical models of flow and heat transfer over louvered fin arrays in compact heat exchangers. Int J Heat Mass Transfer 41:4063–4080

Aziz A (1975) Periodic heat transfer in annular fins. ASME J Heat Transfer 97:302–303

Baek S, Lee C, Jeong S (2014) Effect of flow maldistribution and axial conduction on compact microchannel heat exchanger. Cryogenics 60:49–69

Bhattacharya A, Mahajan R (2000) Finned metal foam heat sinks for electronics cooling in forced convection. In: Proc thirty-fourth national heat transfer conference, Pittsburgh, PA

Burns JC, Parkes T (1967) Peristaltic motion. J Fluid Mech 29:731–743

Chen CK, Chen CH (1990) Nonuniform porosity and non-Darcian effects on conjugate mixed convection heat transfer from a plate fin in porous media. Int J Heat Fluid Flow 11:65–71

Chen TY, Shu HT (2004) Flow structures and heat transfer characteristics in fan flows with and without delta-wing vortex generators. Exp Therm Fluid Sci 28:273–282

Chomdee S, Kiatsiriroat T (2006) Enhancement of air cooling in staggered array of electronic modules by integrating delta winglet vortex generators. Int Commun Heat Mass Transfer 33:618–626

Choudhury C, Garg HP (1991) Design analysis of corrugated and flat plate solar air heaters. Renew Energy 1:595–607

Chu HS, Chen CK, Weng CI (1983) Transient response of circular pins. J Heat Transf 105 (1):205–208

Chyu HK, Hsing YC, Natarajan V (1996) Convective heat transfer of cubic fin arrays in a narrow channel. ASME paper No. 96-GT-201

Chyu MK (1990) Heat transfer and pressure drop for short pin-fin arrays with pin endwall. J Heat Transf 112:926–932

Chyu MK, Goldstein RJ (1991) Influence of an array of wall-mounted cylinders on the mass transfer from a flat surface. Int J Heat Mass Transfer 34(9):2175–2186

Cicfalo M, Stasiek J, Collins MW (1996) Investigation of flow and heat transfer in corrugated passages II numerical results. Int J Heat Mass Transfer 39:165–192

Comini G, Nonino C, Savino S (2003) Effect of aspect ratio on convection enhancement in wavy channels. Numer Heat Transfer Part A Appl 44(1):21–37

Cuta J (1998) Forced convection heat transfer in flat plate microchannel heat exchangers. ASHRAE Trans 104:2

Dake T, Majdalani J (2009) Improving flow circulation in heat sinks using quadrupole vortices. In: Proceedings of the ASME. International PACK conference. American Society of Mechanical Engineers, San Francisco, CA

DeJong NC, Jacobi AM (2003) Localized flow and heat transfer interactions in louvered fin arrays. Int J Heat Mass Transfer 46:443–445

Donahoo EE, Camci C, Kulkarni AK, Belegundu AD (2001) Determination of optimal row spacing for a staggered cross-pin array in a turbine blade cooling passage. J Enhanc Heat Transf 8(1):41–53

Ebisu T (1999) Development of new concept air-cooled heat exchanger for energy conservation of air conditioning machine. In: Kakac S et al (eds) Heat transfer enhancement of heat exchangers. Kluwer Academic, Dordrecht, pp 601–620

Ergin S, Ota M, Yamaguchi H, Sakamoto M (1997) Analysis of periodically fully developed turbulent flow in a corrugated duct using various turbulent models and comparison with experiments. In: JSME centennial grand congress, int conj on fluid eng, Tokyo, Japan, pp 1527–1532

Ferrouillat S, Tochon P, Garnier C, Peerhossaini H (2006) Intensification of heat transfer and mixing in multifunctional heat exchangers by artificially generated streamwise vorticity. Appl Therm Eng 26:1820–1829

Focke WW, Knibbe PG (1986) Flow visualization in parallel plate ducts with corrugated walls. J Fluid Mech 165:73–77

Focke WW, Zachariades J, Oliver I (1985) The effect of the corrugation inclination angle on the thermohydraulic performance of plate heat exchanger. Int J Heat Mass Transfer 28:1469–1479

Fujii M, Seshimo Y, Yarnananaka G (1988) Heat transfer and pressure drop of the perforated surface heat exchanger with passage enlargement and contraction. Int J Heat Mass Transfer 31:135–142

Fujii M, Seshimo Y, Yoshida T (1991) Heat transfer and pressure drop of tube-fin heat exchanger with trapezoidal perforated fins. In: Lloyd JR, Kurosake Y (eds) Proc 1991 ASME-JSME joint thermal engineering conf, vol 4. ASME, NY, pp 355–360

Gaiser G, Kottke V (1990) Effects of corrugation parameters on local and integral heat transfer in plate heat exchangers and regenerators. In: Heat transfer, vol 5. Hemisphere Publishing Corporation, New York, pp 85–90

Goldstein LJ, Sparrow EM (1977) Heat/mass transfer characteristics for flow in a corrugated wall channel. J Heat Transfer 99:187–195

Gong L, Kota K, Tao W, Joshi Y (2011) Parametric numerical study of flow and heat transfer in microchannels with wavy walls. J Heat Transfer 133(5):051702

Gschwind P, Regele A, Kottke V (1995) Sinusoidal wavy channels with Taylor-Gortler vortices. Exp Therm Fluid Sci 11:270–275

Hamaguchi K, Takahashi S, Miyabe H (1983) Heat transfer characteristics of a regenerator matrix (case of packed wire gauzes). Trans JSME 49B(445):2001–2009

Heggs PJ, Walton C (1999) Local heat transfer coefficients in corrugated plate heat exchanger channels with mixed inclination angles. In: IMeChE conf trans 6th UK national conf on heat transfer, pp 39–44

Heikal MR, Drakulic R, Cowell TA (1999) Multi-louvered fin surfaces, in recent advances in analysis of heat transfer for fin type surfaces. In: Sunden B, Heggs PJ (eds) Computational mechanics, Billerica, MA, pp 277–293

Hendricks JB, Nilles MJ, Calkins ME (1995) Heat transfer and flow friction in perforated plate heat exchangers. Exp Therm Fluid Sci 10(2):238–247

Henze M, von Wolfersdorf J (2011) Influence of approach flow conditions on heat transfer behind vortex generators. Int J Heat Mass Transfer 54:279–287

Henze M, von Wolfersdorf J, Weigand B, Dietz CF, Neumann SO (2011) Flow and heat transfer characteristics behind vortex generators – a benchmark dataset. Int J Heat Fluid Flow 32:318–328

Hessami MA (2002) Thermo-hydraulic performance of cross-corrugated plate heat exchangers. In: Heat transfer, proc 12th int heat transfer conf, vol 4, pp 393–398

Holmes MH (1995) Introduction to perturbation methods. Springer-Verlag, New York

Hotani S, Mori K, Maruta T, Suwa A (1977) Performance of a corrugated plate finned coil. Refrig 52:631–640

Huisseune H, T'Joen C, De Jaeger P, Ameel B, De Schampheleire S, De Paepe M (2013a) Performance enhancement of a louvered fin heat exchanger by using delta winglet vortex generators. Int J Heat Mass Transfer 56:475–487

Huisseune H, T'Joen C, De Jaeger P, Ameel B, De Schampheleire S, De Paepe M (2013b) Influence of the louver and delta winglet geometry on the thermal hydraulic performance of a compound heat exchanger. Int J Heat Mass Transfer 57:58–72

Hwang JJ, Su RJ, Tsai BJ (2000) Transient analysis of 2-D cylindrical pin fin with constant base heat flux and tip convective effects. J Enhanc Heat Transf 7(3):201–206

Islamoglu Y, Parmaksizoglu C (2003) The effect of channel height on the enhanced heat transfer characteristics in a corrugated heat exchanger channel. Appl Therm Eng 23(8):979–987

Iwasaki H, Sasaki T, Ishizuka M (1994) Cooling performance of plate fins for multichip modules. In: Intersociety conf on thermal phenomena in electronic systems. IEEE, pp 144–147

Iyengar M, Bar-Cohen A (2002) Least-energy optimization of forced convection plate fin heat sinks. In: Proc the 8th intersociety conf on thermal and thermomechanical phenomena in electronic systems, ITherm 2002, San Diego, CA, pp 792–799

Jang JY, Chen LK (1997) Numerical analysis of heat transfer and fluid flow in a three-dimensional wavy-fin and tube heat exchanger. Int J Heat Mass Transf 40(16):3981–3990

Joardar A, Jacobi AM (2007) A numerical study of flow and heat transfer enhancement using an array of delta-winglet vortex generators in a fin-and-tube heat exchanger. J Heat Transfer 129:1156–1167

Jung J, Hwang G, Baek S, Jeong S, Rowe AM (2012) Partial flow compensation by transverse bypass configuration in multi-channel cryogenic compact heat exchanger. Cryogenics 52:19–26

Kajino M, Hiramatsu M (1987) Research and development of automotive heat exchangers. In: Yang WJ, Mori Y (eds) Heat transfer in high technology and power engineering. Hemisphere, Washington, DC, pp 420–432

Kawamura K, Nakajima T, Matsushima H (2001) Influence of fan setting height on the cooling performance of a plate-fin-type heat sink with microprocessor. Heat Transfer Asian Res 30 (6):512–520

Kays WM, London AL (1984) Compact hear exchangers, 3rd edn. McGraw-Hill, New Year

Kelkar KM, Patankar SV (1987) Numerical prediction of flow and heat transfer in a parallel plate channel with staggered fins. J Heat Transfer 109:25–30

Kelkar KM, Patankar SV (1989) Numerical prediction of heat transfer and fluid flow in rectangular offset fin arrays. Numer Heat Transfer A 15:149–164

Khoshvaght Aliabadi M, Hormozi F, Hosseini Rad E (2014) New correlations for wavy plate-fin heat exchangers: different working fluids. Int J Numer Meth Heat Fluid Flow 24(5):1086–1108

Kim N-H, Byun HW (2012) Effect of inlet configuration on distribution of air-water annular flow in a header of a parallel flow heat exchanger. J Enhanc Heat Transf 19(3):271–292

Kim NH, Kim DY (2010) Two-phase refrigerant distribution in a parallel-flow heat exchanger. J Enhanc Heat Transf 17(1):59–75

Kim SY, Kuznetsov AV (2003) Optimization of pin-fin heat sinks using anisotropic local thermal nonequilibrium porous model in a jet impinging channel. Numer Heat Transfer Part A Appl 44 (8):771–787

Kim SY, Paek JW, Kang BH (2000) Flow and heat transfer correlations for porous fin in a plate-fin heat exchanger. J Heat Transfer 122:572–578

Kirpikov VA, El'chinov VP (1988) The influence of the equivalent diameter on the effectiveness of plate-fin heat-transfer surfaces. Therm Eng 35(2):89–91

Kirsch KL, Thole KA (2017) Heat transfer and pressure loss measurements in additively manufactured wavy microchannels. J Turbomach 139(1):011007

Klett J, Ott R, McMillan A (2000) Heat exchangers for heavy vehicles utilizing high thermal conductivity graphite foams. SAE paper 2000-01-2207, Warrendale, PA

Kouidry F (1997) Etude des Ecoulements Turbuients Charges de Particules: Application a Lencrassment Pa.rticulaire des Echangeurs a Plagues Corruguees. In: PhD thesis. University Joseph Fourier Grenoble, France

Kovalenko VO, Mushin LB, Orlov VK, Isfasman GY, Shevyakova SA (1980) Perforated-plate heat exchangers for cryogenic helium units. Chem Petrol Eng 16(7–8):406–407

Kreid DK, Parry HL, McGowen LJ, Johnson BM (1979) Performance of a plate fin air-cooled heat exchanger with deluged water augmentation. ASME 79-WA/ENER-1

Lau SC, Kim YS, Han JC (1987) Local end wall heat/mass-transfer distributions in pin fin channels. J Thermophys Heat Transfer 1(4):365–372

Lee YN (1986) Heat transfer characteristics of canted-rib plate heat exchanger. In: Heat transfer. Hemisphere Publishing Corporation, Washington, DC, no. 6, pp 2817–2822

Lee PS, Garimella SV, Liu D (2005) Investigation of heat transfer in rectangular microchannels. Int J Heat Mass Transfer 48(9):1688–1704

Lee KS, Kim WS, Si JM (2001) Optimal shape and arrangement of staggered pins in the channel of a plate heat exchanger. Int J Heat Mass Transfer 44:3223–3231

Leu JS, Wu YH, Jang JY (2004) Heat transfer and fluid flow analysis in plate-fin and tube heat exchangers with a pair of block shape vortex generators. Int J Heat Mass Transfer 47:4327–4338

Li HY, Chen KY (2005) Thermal-fluid characteristics of pin-fin heat sinks cooled by impinging jet. J Enhanc Heat Transf 12(2):189–202

Li HY, Chen CL, Chao SM, Liang GF (2013) Enhancing heat transfer in a plate-fin heat sink using delta winglet vortex generators. Int J Heat Mass Transfer 67:666–677

Li HY, Liao WR, Li TY, Chang YZ (2017) Application of vortex generators to heat transfer enhancement of a pin-fin heat sink. Int J Heat Mass Transfer 112:940–949

Li WZ, Yan YY, Shen S, Hai Y (1999) An investigation on heat exchanger performance of a new type of plate heat exchanger with dimples. In: IMeChE conf transactions, 6th UK national conf on heat trans, pp 19–25

Lin WW, Lee DJ (1998) Second-law analysis on wavy plate fin-and-tube heat exchangers. J Heat Transfer 120(3):797–800

Lin WW, Lee DJ (2000) Second-law analysis on a flat plate-fin array under crossflow. Int Commun Heat Mass Transfer 27(2):179–190

Liu MX, Sheu WJ, Chiang SB, Wang CC (2007) PIV investigation of the flow maldistribution in a multi-channel cold plate subject to inlet locations. J Enhanc Heat Transf 14(1):65–76

Loehrke RI, Lane JC, Zelenka RL (1979) Heat transfer from interrupted plate surfaces. Technical Report HT-PP791, Mech Eng Department, Colorado State University

London AL, Shah RK (1968) Offset rectangular plate-fin surfaces heat transfer and flow friction characteristics. ASME J Eng Power 90:218–228

Lu F, Yang J, Kwok DY (2004) Numerical and experimental studies on electrical potential distribution of pressure driven flow in parallel plate microchannels. In: Kandlikar SG (ed) Second international conference on microchannels and minichannels. ASME, ICMM2004-2416

Mahmud S, Islam AS, Mamun MAH (2002) Separation characteristics of fluid flow inside two parallel plates with wavy surface. Int J Eng Sci 40(13):1495–1509

Maiti DK (2002) Heat transfer and flow friction characteristics of plate-fin heat exchanger surfaces – a numerical study, PhD thesis, IIT Kharagpur, India

Majumdar P (2005) Computational methods for heat and mass transfer. Taylor & Francis, New York

Manson SV (1950) Correlations of heat transfer data and of friction data for interrupted plate fins staggered in successive rows, NACA Tech. Note 2237. National Advisory Committee for Aeronautics, Washington, DC

Mao J, Rooke S (1994) Transient analysis of extended surfaces with convective tip. Int Commun Heat Mass Transfer 21(1):85–94

Matsumoto R, Kikkawa S, Senda M, Suzuki M (2000) End wall heat transfer characteristics with a single oblique pin fin. J Enhanc Heat Transfer 7(3):167–184

Maveety JG, Jung HH (2002) Heat transfer from square pin-fin heat sinks using air impingement cooling. IEEE Trans Compon Packag Technol 25(3):459–469

McNab CA, Atkinson KN, Heikal MR, Taylor N (1998) Numerical modeling of heat transfer and fluid flow over herringbone corrugated fins, heat transfer 1998, proceedings of 11. Int Heat Transfer Conf 6:119–124

Mercier P, Tochon P (1997) Analysis of turbulent flow and heat transfer in compact heat exchangers by pseudo direct numerical simulation. In: Shah RK (ed) Compact heat exchangers for process industries. Begell House, New York, pp 223–230

Metwally HM, Manglik RM (2004) Enhanced heat transfer due to curvature- induced lateral vortices in laminar flows in sinusoidal corrugated plate channels. Int J Heat Mass Transfer 47:2283–2292

Metzger DE, Berry RA, Bronson JP (1982) Developing heat transfer in rectangular ducts with staggered arrays of short pin fins. J Heat Transf 104:700–706

Metzger DE, Fan CS, Haley SW (1984) Effects of pin shape and array orientation on heat transfer and pressure loss in pin-fin arrays. J Eng Gas Turbines Power 106:252–257

Metzger DE, Shepard WB, Haley SW (1986, June) Row resolved heat transfer variations in pin-fin arrays including effects of non-uniform arrays and flow convergence. In: ASME 1986 international gas turbine conference and exhibit. American Society of Mechanical Engineers, pp V004T09A015–V004T09A015

Michallon E, Marvillet C, Lebouche M (1996) Thermal and hydraulic characteristics of plate-fin heat exchangers in single phase flow. In: Afgan N, Carvalho MG, Bar-Cohen A, Butterworth D, Roetzel W (eds) New dev in heat exchangers. Gordon and Breach Publishers, London, pp 351–362

Mikulln EI, Shevich YA, Potapov VN (1979) Efficiency of perforated plate array heat exchangers. Chem Petrol Eng 15(5–6):351–355

Min C, Qi C, Kong X, Dong J (2010) Experimental study of rectangular channel with modified rectangular longitudinal vortex generators. Int J Heat Mass Transfer 53:3023–3029

Minakami K, Ishizuka M, Mochizuki S (1995) Performance evaluation of pin-fin heat sinks utilizing a local heating method. J Enhanc Heat Transf 2:17–22

Minakami K, Mochizuki S (1992) Heat transfer characteristics of pin-fins with in-line arrangement (effect of the pin pitch). Proc JSME 920-17:242

Minakami K, Mochizuki S, Murata A, Yagi Y, Iwasaki H (1992) Visualization of flow mixing mechanisms in pin-fin arrays. In: Flow visualization. Springer, Berlin, Heidelberg, pp 504–508

Minakami K, Mochizuki S, Murata A, Yagi Y, Iwasaki H (1993a) Heat transfer characteristics of the pin-fin heat sink (mechanism and effect of turbulence in the pin array). In: International symposium on transport phenomena in thermal engineering, pp 67–72

Minakami K, Mochizuki S, Murata A, Yagi Y, Iwasaki H (1993b) Heat transfer characteristics of pin-fins with in-line arrangement (1st report, effect of the pin pitch). Trans Jpn Soc Mech Eng Part B 59(567):3602–3609

Mithun Krishna PM, Deepu M, Shine SR (2018) Numerical investigation of wavy microchannels with rectangular cross section. J Enhanc Heat Transf 25(4–5):293–313

Mochizuki S, Yagi S (1975) Heat transfer and friction characteristics of strip fins. Int J Refrig 50:36–59

Mochizuki S, Yagi Y, Yang WJ (1988) Flow pattern and turbulent intensity in stacks of interrupted parallel plate surfaces. Exp Therm Fluid Sci I:51–57

Mochizuki S, Yagi Y (1990) Performance evaluation of pin-fin heat exchangers by automated transient testing method. In: Heat Trans 1990, vol 3. Hemisphere Publishing Corporation, pp 217–212

Moffat RJ (1988) Describing the uncertainties in experimental results. Exp Therm Fluid Sci 1:3–17

Mohammed H, Gunnasegaran P, Shuaib N (2011) Numerical simulation of heat transfer enhancement in wavy microchannel heat sink. Int Commun Heat Mass Transfer 38(1):63–68

Molki M, Yuen CM (1986) Effect of interwall spacing on heat transfer and pressure drop in a corrugated wall channel. Int J Heat Mass Transfer 29:987–997

Morini GL (2004) Single-phase convective heat transfer in microchannels: a review of experimental results. Int J Therm Sci 43(7):631–651

Muley A, Borghese JB, White SL, Manglik RM (2006) Enhanced thermal-hydraulic performance of a wavy-plate-fin compact heat exchanger: effect of corrugation severity. In: ASME 2006 international mechanical engineering congress and exposition, pp 701–707

Mullisen RS, Loehrke RI (1983) Enhanced heat transfer in parallel plate arrays. ASME. 83-Ht-43

Naik S, Probert SD, Shilston MJ (1986) Maximizing the performances of flat-plate heat exchangers experiencing free or forced convection. Appl Energy 22:225–239

Nilles MJ, Calkins ME, Dingus ML, Hendricks JB (1995) Heat transfer and flow friction in perforated plate heat exchangers. Exp Therm Fluid Sci 11:238–247

Nunez MP, Polley GT (1999) Methodology for the design of multi-stream plate-fin heat exchangers. In: Sunden B, Heggs PJ (eds) Recent advances in analysis of heat transfer for fin type surfaces. Computational Mechanics, Billerica, MA, pp 277–293

O'Brien JE, Sparrow EM (1982) Corrugated-duct heat transfer, pressure drop and flow visualization. J Heat Transfer 104:410–416

Okada K, Ono M, Tomimura T, Okuma T, Konno H, Ohtani S (1972) Design and heat transfer characteristics of new plate type heat exchanger. Heat Trans Jpn Res 1(1):90–95

Onishi H, Inaoka K, Suzuki K (2001) A three dimensional unsteady numerical analysis for a plate-finned heat exchanger in the middle Reynolds number range. In: Shah RK, Deakin AW, Honda H, Rudy TM (eds) Proc of the third int conf on compact heat exchangers and enhancement technology for the process industries. Begell House Inc., New York, pp 9–16

Oosthuizen PH, Garrett M, (2001) A numerical study of natural convective heat transfer from an inclined plate with a "wavy" surface. In: Proc of 2001 ASME int mech eng congress and exposition. ASME paper no IMECE2001/HTD-24112, New York, NY

Oviedo-Tolentino F, Romero-M'endez R, Hern'andez-Guerrero A, Gir'on-Palomares B (2008) Experimental study of fluid flow in the entrance of a Sinusoidal Channel. Int J Heat Fluid Flow 29(5):1233–1239

Park K, Choi D, Lee K (2004) Optimum design of plate heat exchangers with staggered pin arrays. Numer Heat Transfer Part A 45:347–361

Park K, Rew KH, Kwon JT, Kim BS (2007) Optimal solutions of pin-fin type heat sinks for different fin shapes. J Enhanc Heat Transf 14(2):93–104

Patankar SV, Prakash C (1981) An analysis of the effect of plate thickness on laminar flow and heat transfer in interrupted plate passages. Int J Heat Mass Transfer 24:1801–1810

Peng Y (1983) Heat transfer and friction loss characteristics of pin fin cooling configurations. J Engg. for Gas Turbines and Power 106(1):246–251

Pesteei SM, Subbarao PMV, Agarwal RS (2005) Experimental study of the effect of winglet location on heat transfer enhancement and pressure drop in fin-tube heat exchangers. Appl Therm Eng 25:1684–1696

Picon-Nuñez M, Polley GT, Torres-Reyes E, Gallegos-Muñoz A (1999) Surface selection and design of plate–fin heat exchangers. Appl Therm Eng 19:917–931

Prakash C, Lounsbury R (1986) Analysis of laminar fully developed flow in plate-fin passages: effect of fin shape. J Heat Transfer 108:693–697

Prakash C, Sparrow EM (1980) Natural convection heat transfer performance evaluations for discrete-(in-line or staggered) and continuous-plate arrays. Numer Heat Transfer 3(1):89–105

Qu W, Mudawar I (2002) Experimental and numerical study of pressure drop and heat transfer in a single-phase micro-channel heat sink. Int J Heat Mass Transfer 45(12):2549–2565

Ranganayakulu Ch, Sheik Ismail L, Vengudupathi C (2006) Uncertainties in estimation of Colburn (j) factor and Fanning friction (f) factor for offset strip fin and wavy fin compact heat exchanger surfaces. In: Mishra SC, Prasad BVSSS, Garimella SV (eds) Proceedings of the XVIII national and VII ISHMT – ASME heat and mass transfer conference, Guwahati India, pp 1096–1103

Riddell RA (1966) Heat transfer and flow friction characteristics of a plate-fin type cross-flow heat exchanger with perforated fins. MS thesis in Mech Eng U.S. Naval Postgraduate School

Rocha LAO, Saboya FEM, Vargas JVC (1997) A comparative study of elliptical and circular sections in one- and two-row tubes and plate fin heat exchangers. Int J Heat Fluid Flow 18:247–252

Rodrigues R, Yanagihara JI (1999) Augmentation of heat transfer by longitudinal vortices in plate-fin heat exchanges. In: Proc of the 5th ASME/JSME thermal eng joint conference paper ATJE99-6407

Rodriquez JI, Mills AF (1996) Heat transfer and flow friction characteristics of perforated-plate heat exchangers. Exp Heat Transfer 9(4):335–356

Ros C, Jallut C, Grillot JM, Amblard M (1995) A transient-state technique for the heat transfer coefficient measurement in a corrugated plate heat exchanger channel based on frequency response and residence time distribution. Int J Heat Mass Transfer 38:1317–1325

Rosaguti NR, Fletcher DF, Haynes BS (2006) Laminar flow and heat transfer in a periodic serpentine channel with semi-circular cross-section. Int J Heat Mass Transfer 49(17):2912–2923

Rosaguti NR, Fletcher DF, Haynes BS (2007) Low-Reynolds number heat transfer enhancement in sinusoidal channels. Chem Eng Sci 62(3):694–702

Rosenblad G, Kullendorf A (1975) Estimating heat transfer rates from mass transfer studies on plate heat exchanger surfaces. Wiirme Stoffubertrag 8:187–191

Rosman EC, CaraiLlescov P, Saboya FEM (1984) Performance of one-and two-row tube and plate fin heat exchangers. J Heat Transfer 106:627–632

Rostami J, Abbassi A, Saffar-Avval M (2015) Optimization of conjugate heat transfer in wavy walls microchannels. Appl Therm Eng 82:318–328

Rush T, Newell T, Jacobi A (1999) An experimental study of flow and heat transfer in sinusoidal wavy passages. Int J Heat Mass Transfer 42(9):1541–1553

Saboya FE, Sparrow EM (1976) Transfer characteristics of two-row plate fin and tube heat exchanger configurations. Int J Heat Mass Transfer 19(1):41–49

Saha A (2008) Effect of the number of periodic module on flow and heat transfer in a periodic array of cubic pin-fins inside a channel. J Enhanc Heat Transf 15(3):243–260

Saha AK, Acharya S (2004) Unsteady flow and heat transfer in parallel plate heat exchangers with in-line and staggered arrays of posts. Numer Heat Transfer Part 45:101–133

Saniei N, Dini S (1993) Heat transfer characteristics in a wavy-walled channel. J Heat Transfer 115 (3):788–792

Sasao K, Honma M, Nishihara A, Atarashi T (1999) Numerical analysis of impinging air flow and heat transfer in plate-fin type heat sinks. In: Proc of the Pacific rim/ASME int intersociety electronic and photonic packaging conf, Maui Hawaii, Inter PACK EEP, vol 26, no 1, pp 493–499

Shah RK (1975) Perforated heat exchanger surfaces: prut 2-heat transfer and flow friction characteristics. ASME Paper 75-WA/HT-9, New York

Shah RK, London AL (1978) Laminar flow forced convection in ducts, supplement 1 to advances in heat transfer. Academic Press, New York

Shen J, Gu W, Zhang Y (1987) An investigation on the heat transfer augmentation and friction loss performances of plate-perforated fin surfaces. In: Wang BX (ed) Heat transfer science and technology. Hemisphere, Washington, DC, pp 798–804

Shohtani H (1990) The experimental research on condensation of the plate-fin heat exchangers, using non-azeotropic refrigerant mixtures. In: Proceedings of the second international symposium on condensers and condensation, University of Bath, UK, Heat transfer and fluid flow service, pp 367–376

Shwaish IK, Amon CH, Murthy JY (2002) Thermal/fluid performance evaluation of serrated plate fin heat sinks. In: Proc of the 8th intersociety conf on thermal and thermo-mech phenomena in electronic systems. ITherm, San Diego, CA, pp 267–275

Shwaish IK, Murthy JY, Amon CH, Bains D (2001) Performance evaluation of serrated plate fins for under-carriage electronics cooling in transportation application. In: Proc ASME int mech eng congress and exposition, ASME, New York, paper no IMECE2001/HTD-24391

Sheik Ismail L, Ranganayakulu C, Shah RK (2009) Numerical study of flow patterns of compact plate-fin heat exchangers and generation of design data for offset and wavy fins. Int J Heat Mass Transf 52(17–18):3972–3983

Singh SK, Mishran M, Jha PK (2014) Nonuniformities in compact heat exchangers—scope for better energy utilization: a review. Renew Sustain Energy Rev 40:583–596

Sinha A, Raman KA, Chattopadhyay H, Biswas G (2013) Effects of different orientations of winglet arrays on the performance of plate-fin heat exchangers. Int J Heat Mass Transfer 57:202–214

Song R, Cui M, Liu J (2017) A correlation for heat transfer and flow friction characteristics of the offset strip fin heat exchanger. Int J Heat Mass Transfer 115:695–705

Soodphakdee D, Behnia M, Copeland D (2000) A comparison of fin geometrics for heat sinks in laminar forced convection: Part I – Round elliptical and plate fins in staggered and inline configurations. Int J Microcirc Electron Packag 24(1):68–76

Souidi N, Bontemps A (2001) Concurrent gas-liquid flow in plate-fin heat exchangers with plain and perforated fins. Int J Heat Fluid Flow 22(4):450–459

Sparrow EM, Baliga BR, Patankar SV (1977) Heat transfer and fluid flow analysis of interrupted wall channels with application to heat exchangers. J Heat Transfer 99:4–11

Sparrow EM, Hossfeld M (1984) Effect of rounding protruding edges on heat transfer and pressure drop in a duct. Int J Heat Mass Transfer 27:1715–1723

Stasiek J, Collins MW, Ciofalo M, Chew PE (1996) Investigation of flow and heat transfer in corrugated passages – I. Experimental results. Int J Heat Mass Transfer 39:149–164

Suga T, Aoki H (1991) Numerical study on heat transfer and pressure drop in multilouvered fins. In: Proc ASME/JSME joint thermal engineering conference, vol 4

Suga K, Aoki H (1995) Numerical study on heat transfer and pressure drop in multilouvered fins. J Enhanc Heat Transfer 2(3):231–238

Suga K, Aoki H, Shinagawa T (1990) Numerical analysis on two-dimensional flow and heat transfer of louvered fins using overlaid grids. JSME Int J Ser 2 33(1):122–127

Sui Y, Lee P, Teo C (2011) An experimental study of flow friction and heat transfer in wavy microchannels with rectangular cross section. Int J Therm Sci 50(12):2473–2482

Sui Y, Teo C, Lee P, Chew Y, Shu C (2010) Fluid flow and heat transfer in wavy microchannels. Int J Heat Mass Transfer 53(13–14):2760–2772

Sunden B (1999) Flow and heat transfer mechanisms in plate-frame heat exchangers. In: Kakac S (ed) Heat transfer enhancement of heat exchangers. Kluwer Academic, Dordrecht, pp 185–206

Suryanarayana NV (1975) Transient response of straight fins. J Heat Transfer 97(3):417–423

Suzuki K, Xi GN, Inaoka K, Hagiwara Y (1994) Mechanism of heat transfer enhancement due to self sustained oscillation for an in-line fin array. Int J Heat Mass Transfer 37(1):83–96

Tafti DK, Wang G, Lin W (2000) Flow transition in a multilouvered fin array. Int J Heat Mass Transfer 43:901–919

Tao YB, He YL, Huang J, Wu ZG, Tao WQ (2007a) Numerical study of local heat transfer coefficient and fin efficiency of wavy fin-and-tube heat exchangers. Int J Therm Sci 46:768–778

Tao YB, He YL, Wu ZG, Tao WQ (2007b) Numerical design of an efficient wavy fin surface based on the local heat transfer coefficient study. J Enhanc Heat Transf 14(4):315–332

Theoclitus G (1966) Heat transfer and flow-friction characteristics of nine pin-fin surfaces. J Heat Transfer 88:383–390

Thomas DG (1966) Forced convection mass transfer: Part III. Increased mass transfer from a flat plate caused by the wake from cylinders located near the edge of the boundary layer. AIChE 12:124–130

Tian LT, He YL, Lei YG, Tao WQ (2009) Numerical study of fluid flow and heat transfer in a flat-plate channel with longitudinal vortex generators by applying field synergy principle analysis. Int Commun Heat Mass Transfer 36:111–120

Tischenko ZV, Bondarenko VN (1983) Comparison of the efficiency of smooth-finned plate heat exchangers. Int Chem Eng 23(3):550–557

Tishchenko ZV, Bondarenko VN, Golechek LI (1979) Heat transfer and pressure drop in gas-carrying ducts formed by smooth-finned plate-type heat- exchange surfaces. Heat Trans Sov Res 11(5):117–124

Tolpadi AK, Kuehn TH (1984) Conjugate three-dimensional natural convection heat transfer from a horizontal cylinder with long transverse plate fins. Numer Heat Trans 7:319–141

Torikoshi K, Kawabata K (1989) Heat transfer and flow friction characteristics of a mesh finned air-cooled heat exchanger. In: Figliola RS, Kaviany M, Ebadian MA (eds) Convection heat transfer and transport processes. ASME symp HTD, vol 116, pp 71–77

Toyoshima S, Fukumoto H, Nakagawa Y, Sakamoto Y (1986) Numerical analysis on flow of plate-fin heat exchangers. In: Proceedings 246th lecture meeting of the Japan Society of Mechanical Engineers Kansai Branch, pp 864–871

VanFossen GJ (1981, March) Heat transfer coefficients for staggered arrays of short pin fins. In: ASME 1981 international gas turbine conference and products show. American Society of Mechanical Engineers, pp V003T09A003–V003T09A003

Wang YQ, Dong QW, Liu MS, Wang D (2009) Numerical study on plate-fin heat exchangers with plain fins and serrated fins at low Reynolds number. Chem Eng Technol 32(8):1219–1226

Wang CC, Fu WL, Chang CT (1997) Heat transfer and friction characteristics of typical wavy fin-and-tube heat exchangers. Exp Therm Fluid Sci 14:147–186

Wang CC, Lo J, Lin YT, Wei CS (2002) Flow visualization of annular and delta winglet vortex generators in fin-and-tube heat exchanger application. Int J Heat Mass Transfer 45:3803–3815

Wang G, Vanka S (1995) Convective heat transfer in periodic wavy passages. Int J Heat Mass Transfer 38(17):3219–3230

Webb RL (1987) In: Kakac S, Shah RK, Aung W (eds) Handbook of single-phase heat transfer, vol 17. John Wiley, New York, pp 1–17.62, Chapter 17

Webb RL, Trauger P (1991) The flow structure in the louver fin heat exchanger geometry. Exp Therm Fluid Sci 4:205–217

Wen J, Li K, Wang C, Zhang X, Wang S (2019) Optimization investigation on configuration parameters of sine wavy fin in plate-fin heat exchanger based on fluid structure interaction analysis. Int J Heat Mass Transfer 131:385–402

Wen J, Li Y, Zhou A, Zhang K (2006) An experimental and numerical investigation of flow patterns in the entrance of plate-fin heat exchanger. Int J Heat Mass Transfer 49(9–10):1667–1678

Wen J, Yang HZ, Jian GP, Tong X, Li K, Wang SM (2016a) Energy and cost optimization of shell and tube heat exchanger with helical baffles using Kriging meta model based on MOGA. Int J Heat Mass Transfer 98:29–39

Wen J, Yang H, Tong X, Li K, Wang S, Li Y (2016b) Optimization investigation on configuration parameters of serrated fin in plate-fin heat exchanger using genetic algorithm. Int J Therm Sci 101:116–125

Wen J, Yang H, Tong X, Li K, Wang S, Li Y (2016c) Configuration parameters design and optimization for plate-fin heat exchangers with serrated fin by multi-objective genetic algorithm. Energy Convers Manag 117:482–489

Wieting AR (1975) Empirical correlations for heat transfer and flow friction characteristics of rectangular offset fin heat exchangers. J Heat Transfer 97:488–490

Xi G, Hagiwara Y, Suzuki K (1995) Flow instability and augmented heat transfer of fin arrays. J Enhanc Heat Transf 2:23–32

Xi GN, Shah RK (1999) Numerical analysis of OSF heat transfer and flow friction characteristics. In: Mohamad AA, Sezai I (eds) Proc int conf computational heat and mass transfer. Eastern Mediterranean University Printing house, Eazimaguse, Cyprus, pp 75–87

Xie G, Liu J, Liu Y, Sunden B, Zhang W (2013) Comparative study of thermal performance of longitudinal and transversal-wavy microchannel heat sinks for electronic cooling. J Electron Packag 135:021008–021011

Xu W, Min J (2004) Numerical predictions of fluid flow and heat transfer in corrugated channels. In: Proc intl symp on heat transfer enhancement and energy conservation, Guangzhou, China, vol 1, pp 714–721

Xuan YM, Zhang HL, Ding R (2001) Heat transfer enhancement and flow visualization of wavy perforated plate-and-fin surface. In: Shah RK, Deakin AW, Honda H, Rudy TM (eds) Proc third international conference on compact heat exchangers and enhancement technology for the process industries. Begell House, New York, pp 215–222

Yakushin AN (1977) Determination of optimal characteristics of finned/plate heat exchangers. Therm Eng 24(1):64–67

Yang JW (1972) Periodic heat transfer in straight fins. J Heat Transfer 94(3):310–314

Yang LC, Asako Y, Yamaguchi Y, Faghri M (1995) Numerical prediction of transitional characteristics of flow and heat transfer in a corrugated duct in heat transfer in turbulent flows. ASME Symp Ser HTD 318:145–152

Yang KS, Li SL, Chen IY, Chien KH, Hu R, Wang CC (2010) An experimental investigation of air cooling thermal module using various enhancements at low Reynolds number region. Int J Heat Mass Transfer 53:5675–5681

Yang H, Wen J, Tong X, Li K, Wang S, Li Y (2016) Numerical investigation on configuration improvement of a plate-fin heat exchanger with perforated wing-panel header. J Enhanc Heat Transf 23(1):1–21

Yu X, Feng J, Feng Q, Wang Q (2005) Development of plate-pin fin heat sink and its performance comparisons with a plate fin heat sink. Appl Therm Eng 24:173–182

Yu D, Jeon W, Kim SJ (2017) Analytic solutions of the friction factor and the Nusselt number for the low-Reynolds number flow between two wavy plate fins. Int J Heat Mass Transfer 115:307–316

Zhang LW, Balachandar S, Tafti DK, Najjar FM (1997) Heat transfer enhancement mechanisms in inline and staggered parallel-plate fin heat exchangers. Int J Heat Mass Transfer 40:2307–2325

Zhang J, Kundu J, Manglik RM (2004) Effect of fin waviness and spacing on the lateral vortex structure and laminar heat transfer in wavy-plate-fin cores. Int J Heat Mass Transfer 47:1719–1730

Zhang LW, Tafti DK, Najjar FM, Balachandar S (1997) Computations of flow and heat transfer in parallel plate fin heat exchangers on the CM-5: effects of flow unsteadiness and three-dimensionality. Int J Heat Mass Transfer 40:1325–1341

Zhang Z, Yanzhong L (2003) CFD simulation on inlet configuration of plate-fin heat exchanger. Int J Cryogenic Eng 43:673–678

Zhou J, Hatami M, Song D, Jing D (2016) Design of microchannel heat sink with wavy channel and its time-efficient optimization with combined RSM and FVM methods. Int J Heat Mass Transfer 103:715–724

Chapter 6
Conclusions

The details of plate-fin surface geometries have been presented in this monograph. Different designs of plate-fins used for compact heat exchangers have also been reported. The applications of plate-fin extended surfaces for heat transfer enhancement in both single-phase and two-phase flows, as studied by researchers across the globe, have been discussed. The works carried out on the fouling of compact heat exchangers have also been dealt with.

The following chapter intensively deals with offset-strip fins for plate-fin heat exchangers and plate-fin and tube heat exchangers. The performance of offset-strip fins and offset-strip fin arrays has been discussed. Various numerical studies and correlations for j factor and f factor have been presented.

The concepts of louvered fins and convex louvered fins have been briefed in the next chapter. The effect of fin parameters on heat transfer and pressure drop characteristics has been discussed. The performance comparison of louvered fins with that of offset-strip fins has also been presented. The correlations developed for louvered fins by different researchers have been considered.

Chapter 4 has been dedicated to understanding the basic concepts of vortex generators for heat transfer enhancement in plate-fin heat exchangers. The performance of transverse, longitudinal, and wing-type vortex generators has been discussed.

The performance of different plate-fins such as wavy fins, corrugated fins, perforated fins, pin fins, pin-fin arrays, wire meshes, metal foam fins, and packing has been presented in Chap. 5. The use of plate-fins for cooling heat sinks has also been discussed. The numerical studies on heat transfer enhancement performance of plate-fins have been mentioned.

Chapter 6 concludes the research monograph.

© The Author(s), under exclusive license to Springer Nature Switzerland AG 2020   137
S. K. Saha et al., *Heat Transfer Enhancement in Plate and Fin Extended Surfaces*,
SpringerBriefs in Applied Sciences and Technology,
https://doi.org/10.1007/978-3-030-20736-6_6

Additional References

Armstrong JE, Winstanley DA (1988) A review of staggered array pin fin heat transfer for turbine cooling applications. J Turbomach 110(1):94–103

Brockmeier U, Feibig M, Guntermann T (1989) Heat transfer enhancement in fin-plate heat exchangers by wing type vortex generators. Chem Eng Technol 12(4):288–294

Burgers JG, Lemczyk TF (1988) Optimization of louvered fins in intermittent contact with plate heat exchanger passageways. In: SAE int congress and exposition, SAE 880447

Chyu MK, Schwarz SG (1990) Effects of bottom injection on heat transfer and fluid flow in rectangular cavities. J Thermophys Heat Transfer 4(4):521–526

Chyu MK, Hsing YC, Natarajan V (1998) Convective heat transfer of cubic fin arrays in a narrow channel. J Turbomach 120(2):362–367

DeJong NC, Jacobi AM (1999) Local flow and heat transfer behavior in convex-louver fin arrays. J Heat Transfer 121:136–141

Dubrovsky EV, Fedotova AJ (1972) Investigation of heat-exchanger surfaces with plate fins. Heat Transfer Sov Res 4(6):75–79

Edwards FJ, Alker CJ, Crompton N (1974) FC6. 4 The improvement of forced convection surface heat transfer using surface protrusions in the form of (a) cubes and (b) vortex generators, Digital Library, IHTC, https://doi.org/10.1615/IHTC5.2140, pp 244–248

Fiebig M (1995) Vortex generators for compact heat exchangers. J Enhanc Heat Transf 2:43–62

Han DH, Lee KJ, Kim YH (2003) Experiments on the characteristics of evaporation of R410A in brazed plate heat exchangers with different geometric configurations. Appl Therm Eng 23:1209–1225

Ho AMS, Qu W, Pfefferkorn F (2006) Pressure drop and heat transfer in a single-phase micro-pin-fin heat sink. In: ASME 2006 international mechanical engineering congress and exposition. American Society of Mechanical Engineers, pp 213–220

Hu S, Herold KE (1995a) Prandtl number effect on offset fin heat exchanger performance: experimental results. Int J Heat Mass Transfer 38:1053–1062

Hu S, Herold KE (1995b) Prandtl number effect on offset fin heat exchanger performance: predictive model for heat transfer and pressure drop. Int J Heat Mass Transfer 38:1043–1051

Huihua Z, Xuesheng L (1989) The experimental investigation of oblique angles and interrupted plate lengths for louvered fins in compact heat exchangers. Exp Therm Fluid Sci 2:100–106

Jeon CD, Lee J (2001) Local heat transfer characteristics of louvered plate fin surfaces. ASHRAE Trans 107(1):338–345

Kalinin EK, Dreitser GA, Dubrovsky EV (1985) Compact tube and plate-finned heat exchangers. Heat Transfer Eng 6(1):44–51

© The Author(s), under exclusive license to Springer Nature Switzerland AG 2020   139
S. K. Saha et al., *Heat Transfer Enhancement in Plate and Fin Extended Surfaces*,
SpringerBriefs in Applied Sciences and Technology,
https://doi.org/10.1007/978-3-030-20736-6

Kim JY, Song TH (2002) Microscopic phenomena and macroscopic evaluation of heat transfer from plate fins/circular tube assembly using naphthalene sublimation technique. Int J Heat Mass Transfer 45:3397–3404

Kim JY, Song TH (2003) Effect of tube alignment on the heat/mass transfer from a plate fin and two-tube assembly: naphthalene sublimation results. Int J Heat Mass Transfer 46:3051–3059

Lee KS, Kim WS, Lee TH (1994) An experimental study on the behavior of frost formation in the vertical plate heat exchanger. In: Tree DR, Braun JE (eds) Proc int refrigeration conf at Purdue, pp 329–334

Lee TH, Lee KS, Kim WS (1996) The effects of frost formation in a flat plate finned-tube heat exchanger. In: Braun JE, Groll EA (eds) Proc the 1996 int refrigeration conf at Purdue, pp 205–210

Lee YB, Ro ST (2002) Frost formation on a vertical plate in simultaneously developing flow. Exp Therm Fluid Sci 26:939–945

Matsushima H, Uchida M (2003) Evaporation performance of a plate heat exchanger embossed with pyramid-like structures. J Enhanc Heat Transf 10:171–180

Metzger DE, Haley SW (1982) Heat transfer experiments and flow visualization for arrays of short pin fins. In: ASME 1982 international gas turbine conference and exhibit. American Society of Mechanical Engineers, pp V004T09A007–V004T09A007

Metzger DE, Shepard WB, Haley SW (1986) Row resolved heat transfer variations in pin-fin arrays including effects of non-uniform arrays and flow convergence. In: ASME international gas turbine conference and exhibit. American Society of Mechanical Engineers, pp V004T09A015–V004T09A015

Mitra NK, Fiebig M (1993) Experimental investigations of heat transfer enhancement and flow losses in a channel with double rows of longitudinal vortex generators. Int J Heat Mass Transfer 36(9):2327–2337

Montgomery SR, Wilbulswas P (1967) Laminar flow heat transfer for simultaneously developing velocity and temperature profiles in ducts of rectangular cross-section. Appl Sci Res 18:247–259

Muzychka YS, Yovanovich MM (2001a) Modeling the f and j characteristics of the offset-strip fin array. J Enhanc Heat Transf 8:261–278

Muzychka YS, Yovanovich MM (2001b) Modeling the f and j characteristics for transverse flow through an offset-strip fin at low Reynolds number. J Enhanc Heat Transf 8:243–260

Osada H, Aoki H, Ohara T, Kuroyanagi I (2001) Research on a plate with multi-louvered fins under dehumidification. Heat Trans Asian Res 30(5):383–393

Pescod D (1980) An advance in plate heat exchanger geometry giving increased heat transfer in compact heat exchangers. In: History technological advancement and mechanical design problems, vol 10. ASME, New York, pp 73–77

Prasad BS, Gurukul SM (1992) Differential methods for the performance prediction of multistream plate-fin heat exchangers. J Heat Transfer 114(1):41–49

Rich DG (1973) The effects of fin spacing on the heat transfer and friction performance of multi-row, smooth plate fin-and-tube heat exchangers. ASHRAE Trans 79(2):137–145

Riemann F, Leiner M (1993) Turbulent local heat transfer coefficient in channels with rectangular vortex generator, pp 139–146

Robertson JM (1980) Review of boiling, condensing and other aspects of two- phase flow in plate fin heat exchangers. In: Shah RK, CF MD, Howard CP (eds) Compact heat exchangers-history, technological advances and mech design problems HTD, vol 10. ASME, New York, pp 17–27

Rong-Hua Y (1997) An analytical study of the optimum dimensions of rectangular fins and cylindrical pin fins. Int J Heat Mass Transfer 40(15):3607–3615

Sunden B, Svantesson J (1992) Correlation of j and f factors for multilouvered heat transfer surfaces. In: Proceedings of the 3'" UK national heat transfer conference, pp 805–811

Sundén LW (2001) Design methodology for multistream plate-fin heat exchangers in heat exchanger networks. Heat Transfer Eng 22(6):3–11

Tian L, He Y, Tao Y, Tao W (2009) A comparative study on the air-side performance of wavy fin-and-tube heat exchanger with punched delta winglets in staggered and in-line arrangements. Int J Therm Sci 48(9):1765–1776

VanFossen GJ (1981) Heat transfer coefficients for staggered arrays of short pin fins. In: ASME 1981 international gas turbine conference and products show. American Society of Mechanical Engineers, pp V003T09A003–V003T09A003

Wang SM, Li YZ, Wen J, Ma YS (2010) Experimental investigation of header configuration on two-phase flow distribution in plate-fin heat exchanger. Int Commun Heat Mass Transfer 37:116–120

Webb RL (1983) Enhancement for extended surface geometries used in air-cooled heat exchangers in low Reynolds number flow heat exchangers. Hemisphere, Washington, DC, pp 721–734

Webb RL, Kim NH (2004) Principles enhanced heat transfer. Garland Science, London

Index